SELECTED PIPING PROBLEMS

By

S.D. BOWMAN

AUTHOR OF

"MODERN METHODS OF PIPE FABRICATION"
AND
"ORDINATES FOR 1000 PIPE INTERSECTIONS"

TWELFTH EDITION

2007

TWELFTH EDITION

2007 Printing

The Author: S.D. Bowman
PO BOX 86651
Baton Rouge, LA 70879
Tel: (225) 937-2965
email: stepbow1@att.net

The publisher: CLAITOR'S PUBLISHING DIVISION
3653 Perkins Rd., P.O. Box 261333
Baton Rouge, LA 70826-1333
Tel: 800-274-1403 (In LA 225-344-0476)
Fax: 225-344-0480
email: claitors@claitors.com
Order Form/URL: www.claitors.com

TABLE OF CONTENTS

SELECTED PIPING PROBLEMS

The offsets and intersections in this book represent a number of problems that the pipefitter will be called upon to solve at one time or another during his career.

Since these are typical problems, they can be used as a guide to solve similar problems. They are also suggestive of ways to work unusual or unique problems that the pipefitter might encounter.

Almost all piping problems are worked by solving right triangles. For that reason the formulas that are used to solve these triangles, and the trigonometry tables that go with them are included in this book. The ability to use these formulas and tables are all that is required to work the problems.

FORMULAS FOR FINDING THE DEGREE OF ACUTE ANGLES WHEN TWO SIDES OF THE TRIANGLE ARE KNOWN

$$\frac{\text{Side Opposite}}{\text{Hypotenuse}} = \text{Sine}$$

$$\frac{\text{Side Adjacent}}{\text{Hypotenuse}} = \text{Cosine}$$

$$\frac{\text{Side Opposite}}{\text{Side Adjacent}} = \text{Tangent}$$

$$\frac{\text{Side Adjacent}}{\text{Side Opposite}} = \text{Cotangent}$$

$$\frac{\text{Hypotenuse}}{\text{Side Adjacent}} = \text{Secant}$$

$$\frac{\text{Hypotenuse}}{\text{Side Opposite}} = \text{Cosecant}$$

FORMULAS FOR FINDING THE LENGTH OF SIDES FOR RIGHT TRIANGLES WHEN AN ANGLE AND ONE SIDE ARE KNOWN

Length of side opposite = Hypotenuse X Sine
or
Length of side opposite = Side adjacent X tangent

Length of side adjacent = Hypotenuse X Cosine
or
Length of side adjacent = Side opposite X Cotangent

Length of hypotenuse = Side opposite X Cosecant
or
Length of hypotenuse = Side adjacent X Secant

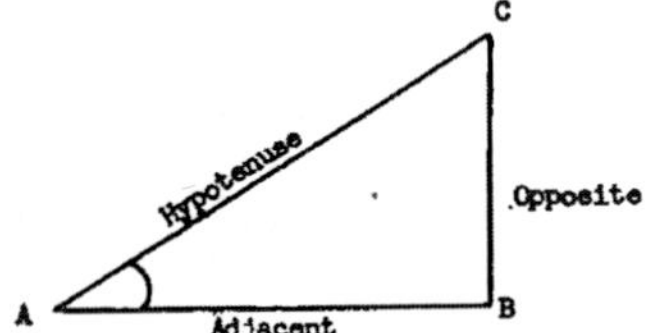

Names of sides of right triangle in relation to angle A

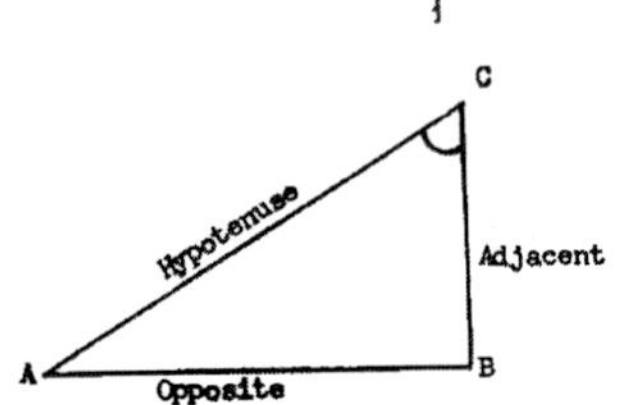

Names of sides of right triangle in relation to angle C

<u>OFFSET WITH DEGREE OF ELBOW UNKNOWN</u>

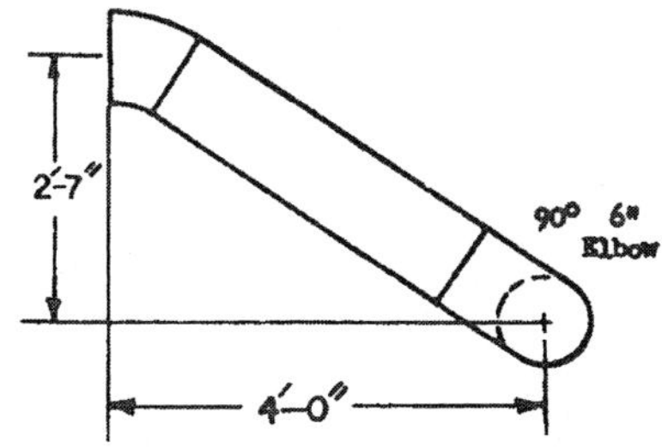

Object: To find the degree of the smaller elbow that will make a 2'-7" offset when used with a 90 degree elbow. Notice that the overall dimension of 4'-0" is from the center of the 90° elbow to the <u>end</u> of the smaller elbow, and the degree of the smaller elbow is not given.

The first step in the solution is to draw the radius line of the small elbow, BD, and the horizontal line AC through the center of the 90° elbow.

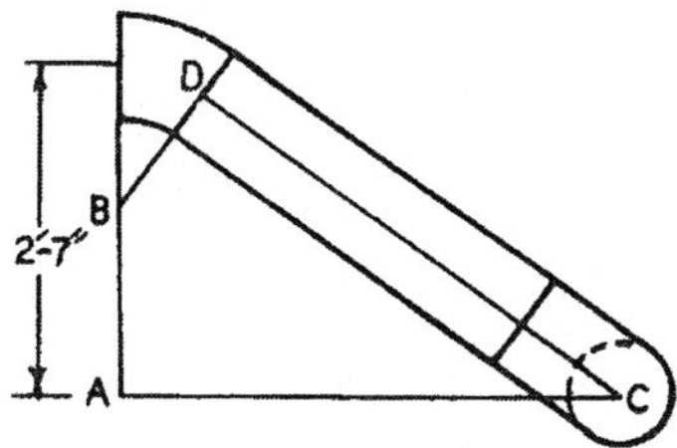

In this example problem the elbows are size 6" so their radius is 9".

Next draw connecting line BC to form triangle BDC and triangle BAC.

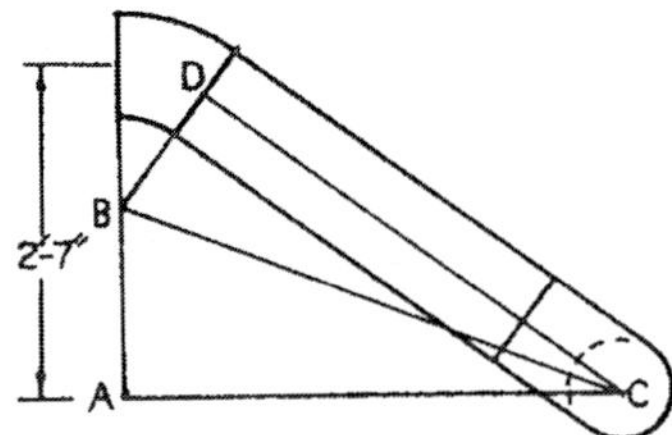

Dimension the triangles. Side AC is 48", the overall dimension. BD is 9", the radius of the 6" elbows. Side AB is found by subtracting the radius of the elbow (9") from the amount of offset (31"). Side AB then is 22".

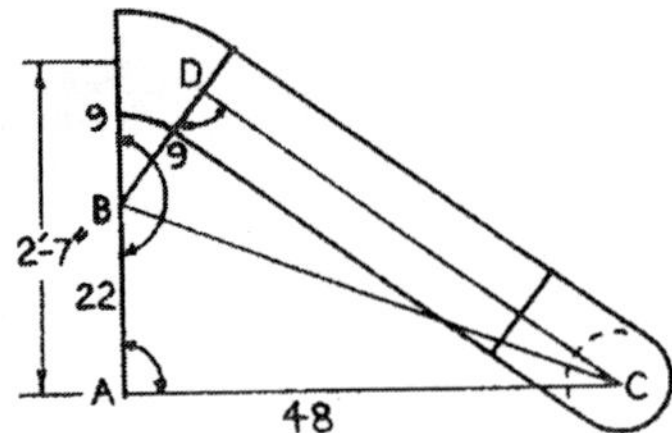

Angle A and angle D are 90 degree angles. It can be seen above that if angle B of triangle BAC and angle B of triangle BDC are found and their sum subtracted from 180 degrees (a straight line), the remaining angle will be the degree of the elbow.

In this example problem:

$$\text{Angle B of triangle BAC} = \frac{\text{Side opposite}}{\text{Side adjacent}}$$

Angle B of triangle BAC = $\frac{48}{22}$ = 2.1818 (Look up under tangent)

Angle B of triangle BAC = 65°-22'

Before finding angle B of triangle BDC, side BC must first be found. Side BC is the hypotenuse of triangle BAC.

Hypotenuse (BC) = Cosecant of angle B times opposite side

BC = Cosecant of 65°-22'X 48

BC = 1.1001 X 48

BC = 52.80"

To find angle B of triangle BDC, divide the hypotenuse by the adjacent side of angle B and look up under secant.

Angle B = $\frac{52.80}{9}$ =5.8672 (Look up under secant)

Angle B = 80"-11'

Add angle B of triangle BAC and angle B of triangle BDC together and subtract their sum from 180° to obtain the degree of the smaller elbow.

65°-22'	179°-60'
+80°-11'	-145°-33'
145°-33'	34°-27'

It would take a 34°-27' elbow to make the offset in this problem.

OFFSET WITH DEGREE OF ELBOWS UNKNOWN

One of the dimensions of this offset is measured from face to face of flanges and for that reason the degree of the elbow cannot be found in the usual way. However, by using the following procedure any offset of this type can be solved. The object is to find the degree of the elbow to use so that the offset will have the dimensions shown below.

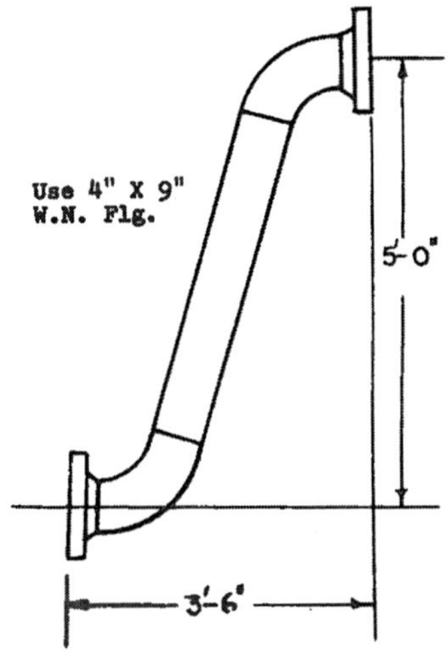

The solution to this problem that follows can be used as a guide in working similar problems.

The first step is to draw in the radius lines of both elbows. These are 4" elbows so their radius is 6".

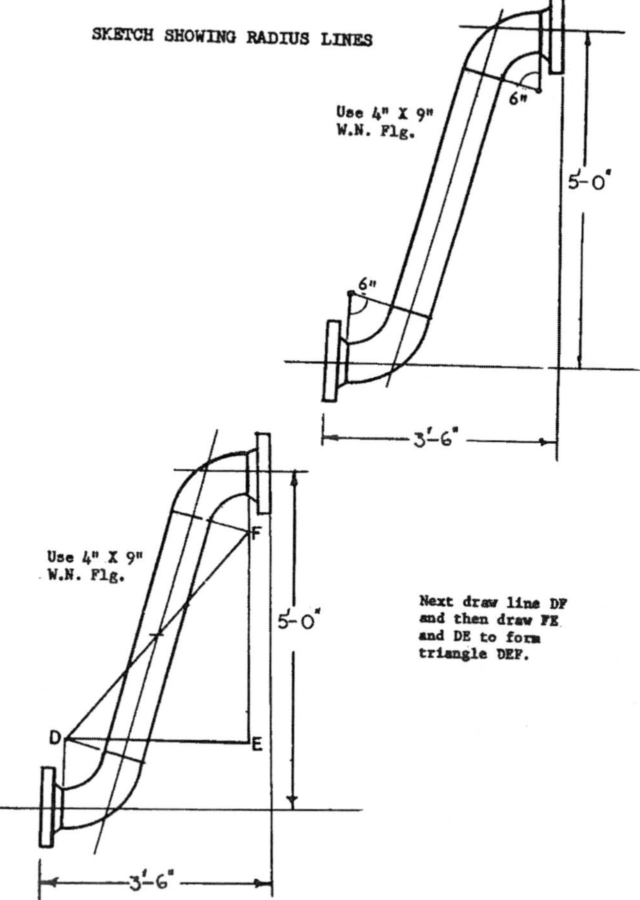

Next draw line DF and then draw FE and DE to form triangle DEF.

Draw line CJ to form right triangle CJF. To solve this problem find angle F of triangle DEF and angle F of triangle CJF. Add these two angles together and subtract their sum from 180°. The remaining angle will be the degree of the elbows.

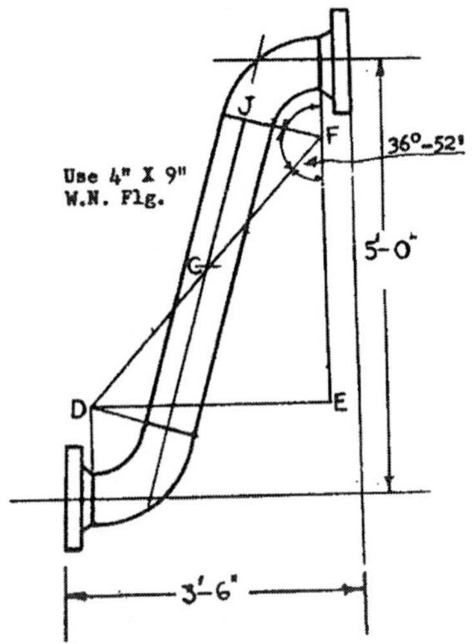

In this example problem side DE of triangle DEF is 36". It is found by subtracting the take up of the two welding neck flanges from the overall dimension 3'-6". Side FE of triangle DEF is 5'-0" minus the radius of each of the two 4" elbows. Side FE then is 48" long.

To find angle F of triangle DEF:

$$\frac{\text{Adjacent side}}{\text{Opposite side}} = \text{Cotangent} \qquad \frac{48}{36} = 1.333$$

The angle that has 1.333 for its cotangent is 36°-52'. This is angle F of triangle DEF.

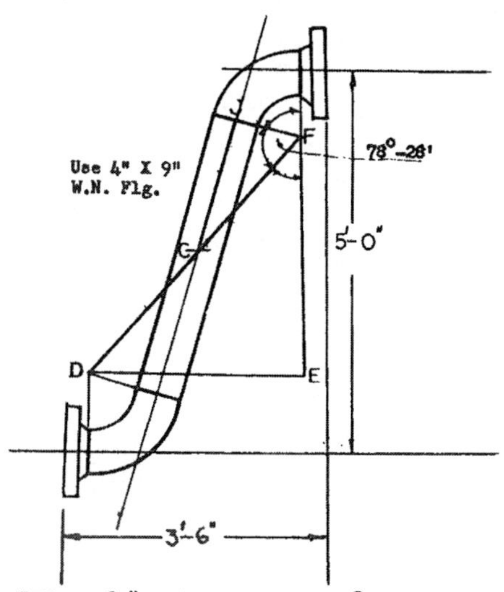

Side FD of triangle DEF is 60". (Cosecant of 36°-52' times 36). Side FC of triangle CJF is 30". (½ of side FD). Side JF of triangle CJF is 6". (radius of 4" elbow).

To find angle F of triangle CJF use the formula:

$\frac{\text{Hypotenuse}}{\text{Adjacent}}$ = Secant $\qquad \frac{30}{6}$ = 5.000 (Secant)

The angle that has 5.000 for its secant is 78°-28'. This is angle F of triangle CJF.

36°-52'
78°-28'
114°-80' = 115°-20' (sum of the two angles)

179°-60'
-115°-20'
64°-40' = the degree of the elbows

OFFSET WITH DEGREE OF ELBOWS UNKNOWN

This is the same problem as the preceding one, but is solved in a slightly different manner.

Given: 5'-0" offset with 3'-6" face to face dimension.

Object: To find degree of elbows to use.

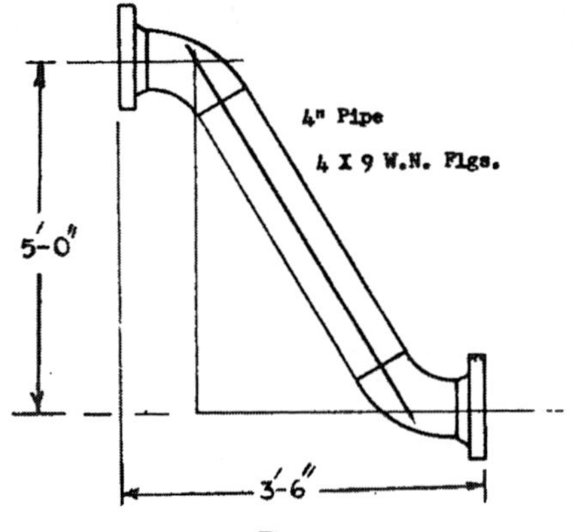

The first step is to subtract the take up of the two flanges from the overall dimension, making the new measurement 3'-0". Next, divide the offset into equal parts with line AB. Line AB will be 18" long and line BE will be 30" long.

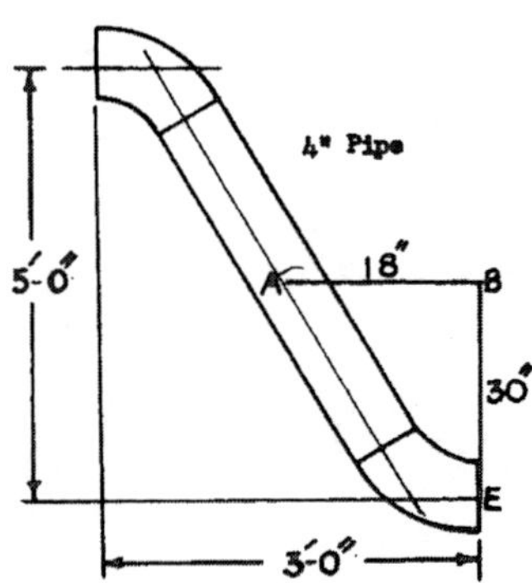

Draw the radius lines EC and DC and the connecting line AC to form triangles ABC and ADC. Notice in the sketch below that these two right triangles have a common hypotenuse, AC.

This problem is solved by finding angle C of triangle ABC and angle C of triangle ADC. Add these two angles together and subtract their sum from 180°. The remaining degree will be the degree of the fitting.

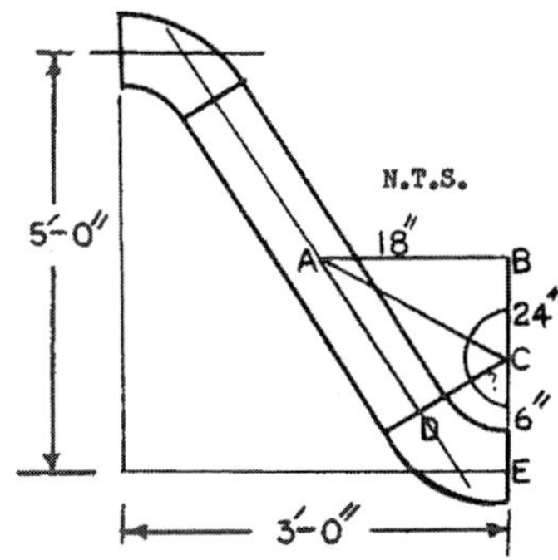

In triangle ABC: Angle C = $\frac{\text{Adjacent side}}{\text{Opposite side}}$

Angle C = $\frac{24}{18}$ = 1.333 (Cotangent)

Angle C = 36°-52'

Side AC equals cosecant of 36°-52' times opposite side (18").

Side AC = 1.6668 X 18

Side AC = 30"....the hypotenuse of both triangles.

In triangle ADC: Angle C = $\frac{\text{Hypotenuse}}{\text{Adjacent side}}$

Angle C = $\frac{30}{6}$ = 5.000 (Secant)

Angle C = 78°-28'

36°-52'
+78°-28'
115°-20' To be subtracted from 180°.

179°-60'
115°-20'
54°-40' Degree of elbows.

HOW TO FIND LENGTH OF RUN PIPE OF ROLLING OFFSET USING 90° ELBOWS

A rolling offset is a double offset, one end of which is rotated or rolled so as to effect a change in elevation and lateral position at the same time. The fittings normally used to make these offsets are 90° or 45° welding elbows. However, any degree fitting can be used depending upon the design of the particular installation.

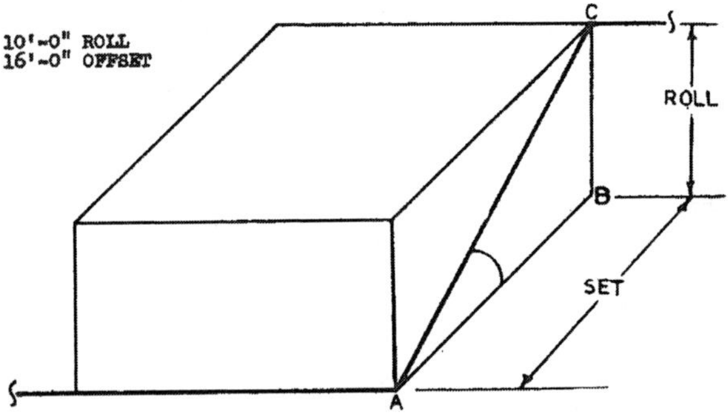

The object is to find the length of the run pipe, which is side AC of the right triangle ABC (above).

It is necessary to find Angle A of triangle ABC before side AC can be found. Side AB is the adjacent side of Angle A in triangle ABC and BC is the opposite side. Use the formula for solving right triangles when two sides and one angle are known. It is:

$$\frac{\text{Side Opposite}}{\text{Side Adjacent}} = \text{Tangent}$$

Substituting in the formula:

$$\frac{10}{16} = .6250 \text{ (Look up under tangent)}$$

From the trig tables the angle that has .6250 for its tangent is 32°-0". This is angle A of triangle ABC.

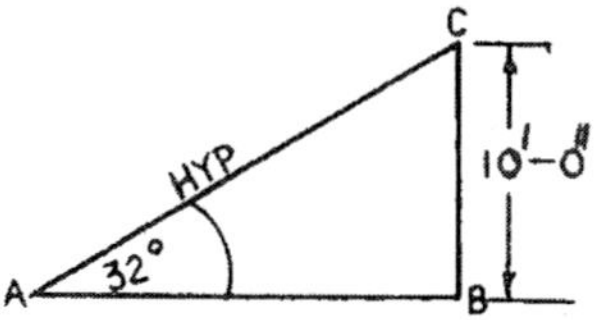

Now that two angles are known (angle A and the right angle) the run pipe, side AC can be found. Use the formula for solving a right triangle when one side and two angles are known:

It is:

Length of hypotenuse = Opposite X Cosecant

Length of hypotenuse = 120" X 1.8871

Hypotenuse (Side AC) = 226.45" or 18'-10 7/16"

The center to center length of the run pipe is 18'-10 7/16". If fittings are used for the 90° turns, subtract their take up from this dimension to obtain the cut length of the pipe.

SOLUTION OF 45° ROLLING OFFSET

Example problem:

1'-0" Roll
1'-8" Offset

The triangles involved in the solution of the rolling offset are easily seen by placing the offset in a box as shown below. The object is to find side AC of triangle ABC, which is the pipe between the two bends.

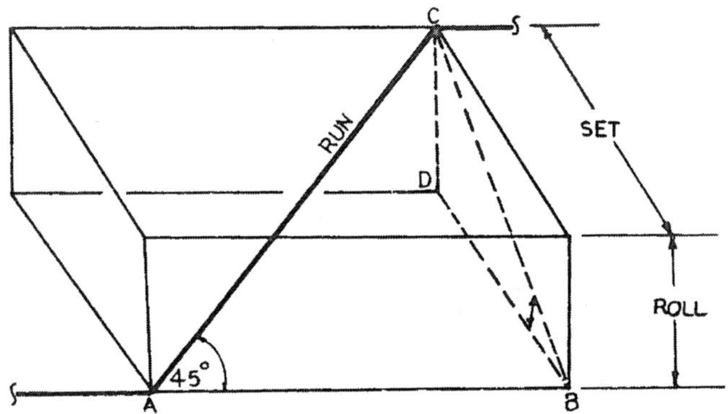

In determining the length of side AC, two triangles must be solved. Side BC triangle BDC is the true offset of the rolling offset. It's length must be found so that it can be multiplied by the cosecant of angle A to find side AC of triangle ABC.

It is necessary to find angle B of triangle BDC before side BC can be found. Side BD is the adjacent side and DC is the opposite side of angle B. Find angle B by using the formula for solving a right triangle when two sides and the right angle is known. It is:

$$\frac{\text{Side Adjacent}}{\text{Side Opposite}} = \text{Cotangent}$$

Substituting in the formula:

$\frac{20}{12}$ = 1.666 (Look up under cotangent in trig tables).

The angle that has 1.666 for its cotangent is 30° - 58' and is angle B of triangle BDC.

Now that angle B is known, side BC in triangle BDC can be found. Use the formula for finding the hypotenuse when the opposite side is known. It is:

Length of hypotenuse = side opposite X cosecant of 30° - 58'.

Substituting in the formula:

Length of hypotenuse = 12 X 1.9435

Length of hypotenuse = 23.32" (side BC)

Side AC can now be found by using the formula for finding the hypotenuse when the opposite side and two angles of triangle ABC are known. It is:

Length of hypotenuse = cosecant times opposite side

Length of hypotenuse = 1.414 X 23.32

Length of hypotenuse = 33" which is center to center length between the two bends. (Side AC of triangle ABC)

HOW TO FIND DEGREE TO SET FLANGE BOLT HOLES

If the rolling offset was made and installed with the dimensions given, two of the flange bolt holes should be level (set square). Notice the bolt holes in the flange in the sketch below.

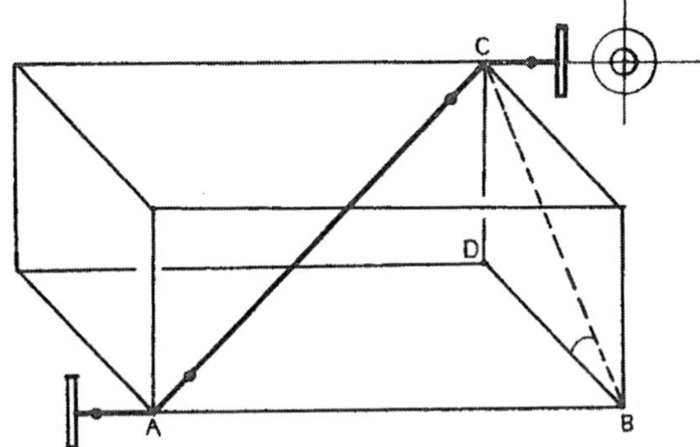

However, if the same offset was lying flat, in the true offset position shown below, two holes would not likely be square, but would be "off" the same number of degrees as contained in angle B of triangle BDC above.

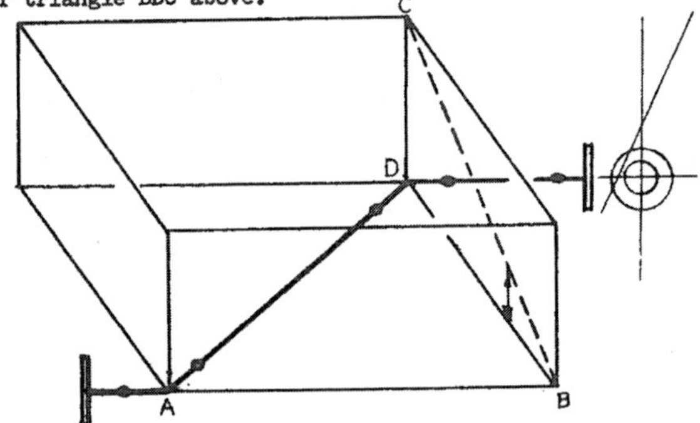

The only time the bolt holes could be set square with the offset lying flat would be if the roll and set dimensions were the same and the flange had 8 or 16 bolt holes.

In solving the rolling offset, angle B of triangle BDC was found to be 30°-58' degrees. With the offset lying flat, the flange bolt holes would be set on this degree past makeup as shown.

The flange bolt holes on the other end of the offset would BE SET to correspond with them. If you were facing the open end of the flanges on the other end, then the bolt holes would be set 30°-58' degrees less than makeup. Angle B will always be the degree to set the holes on.

ROLLING OFFSET--OTHER THAN 45° ELBOWS

The design of some installations require a rolling offset where the travel (also called advance) of the offset is limited and 45 degree elbows cannot be used.

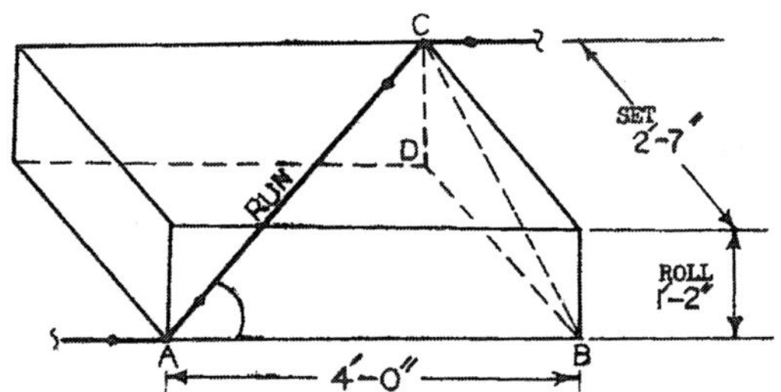

Example problem:

2'-7" Offset
1'-2" Roll
4'-0" Travel

The only difference between this offset and one made with 45° elbows is that after side BC (the true offset) is found, its relation to side AB is found to determine angle A. Angle A of triangle ABC is the degree of the elbows that are required. The cosecant of angle A is then multiplied by side BC to find the length of AC. Side AC is the center to center measurement between the two elbows.

The steps performed to solve this problem are as follows:

Find angle B of triangle BDC
Find side BC of triangle BDC
Find angle A of triangle ABC
Find side AC, the run pipe

It is necessary to find angle B of triangle BDC before side BC can be found. Side BD is the adjacent side and DC is the opposite side when referring to angle B. Find angle B by using the formula for solving a right triangle when two sides and one angle are known.
It is:

$$\frac{\text{Side opposite}}{\text{Side adjacent}} = \text{Tangent}$$

Substituting in the formula:

$\frac{14}{31}$ = .4516 (Look up under tangent in trig table)

The angle that has .4516 for its tangent is 24°-18' and is angle B.

Now that angle B is known, side BC of triangle BDC can be found. Use the formula for finding the hypotenuse when the opposite side is known. It is:

Length of hypotenuse = side opposite times cosecant (of 24°-18')

Length of hypotenuse = 14 X 2.4300

Length of hypotenuse = 34.02" or 2'-10" which is side BC

The degree of angle A of triangle ABC must be determined before side AC can be found. Side AB (adjacent side) is 4'-0" and side BC the opposite side has been found to be 2'10". Use the formula for finding an angle when two sides and one angle are known. It is:

$$\frac{\text{Side adjacent}}{\text{Side opposite}} = \text{Cotangent}$$

Substituting in the formula:

$\frac{48}{34.02}$ = 1.4109 (Look up under cotangent)

From the trig table, the angle that has 1.4109 for its cotangent is 35° -20'. This is angle A of triangle ABC and is the degree of the elbows required to make the rolling offset to the dimensions specified.

Now that angle A is known, side AC of triangle ABC can be found, Side BC is the opposite side and has been found to be 34.02". Use the formula for finding the hypotenuse when one side and two angles are known: It is:

Length of hypotenuse = side opposite X cosecant

Substituting in the formula:

Hypotenuse = 34.02 X 1.7291
Hypotenuse = 58.81" length of run pipe AC

The cut length of the pipe between the two elbows is found by subtracting the take up of the two 35°-20' elbows from the center to center length of the run pipe.

HOW TO FIND DEGREE OF ELBOWS REQUIRED TO MAKE OFFSET THAT RISES AND TURNS AT THE SAME TIME

This type of offset is different from a rolling offset in that a certain degree of rise and degree of turn is specified and the dimensions have to conform to these degrees. In a rolling offset, the amount of offset and roll is given and the degree of rise and turn is not considered. Even in a rolling offset in which the travel (or advance) is specified, the degree of rise and turn is not calculated.

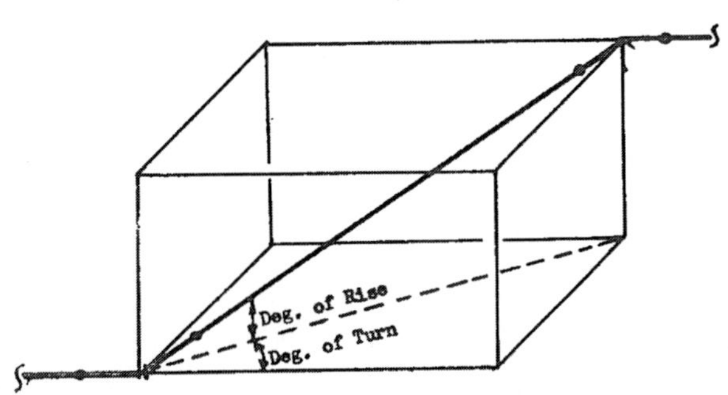

In offsets like the one above, the degree of both the elbows is the same. They are found by multiplying the cosine of the degree of turn times the cosine of the degree of rise. This product is looked up under cosine in the trig tables and its corresponding degree is the degree of the fitting. This is called cosine times cosine equals cosine formula and is used to solve several similar problems..

EXAMPLE PROBLEM: Find degree of elbow that has a degree of rise of 40° and an angle of turn of 25°.

SOLUTION:

Cosine of degree of rise = .766
Cosine of degree of turn = X.906
4596
.6894
.693996

From the trig tables, the degree that has .69399 for its cosine is 46°-54'. This is the degree of the elbows required to make an offset that has a rise of 40° and that makes a turn of 25° at the same time.

The degree of the top elbow is the same as the bottom elbow because the angle of turn for both is the same. (They are alternate angles). The degree of rise is always the same therefore the solution is the same.

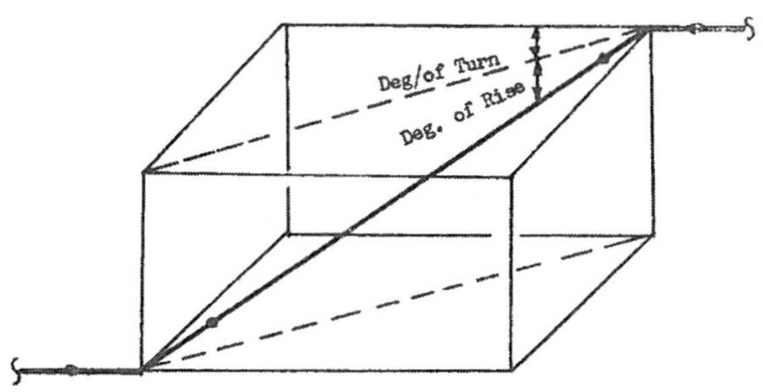

HOW TO FIND DEGREE OF ELBOWS REQUIRED TO MAKE OFFSET THAT RISES AND TURNS AT THE SAME TIME

Example problem: Find degree of elbows required to make an offset that has a rise of 45° and a turn of 45° at the same time as illustrated below.

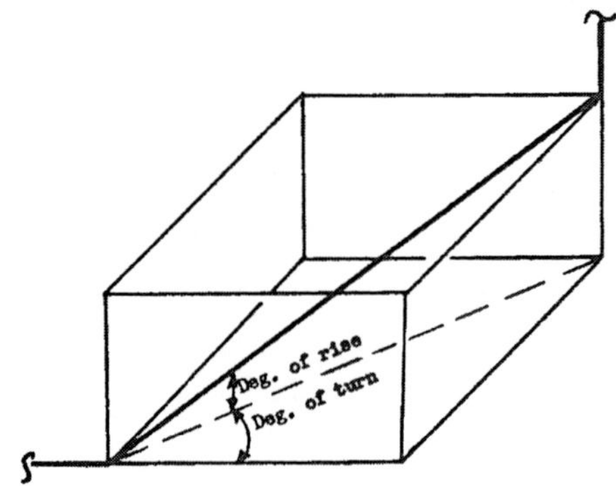

Explanation: In this problem, one end of the bottom elbow is elevated so that the pipe rises as it turns. The degree of the top and bottom elbow is not always the same and must be found separately. The degree of the bottom elbow is found by multiplying the cosine of the angle of the turn times the cosine of the degree of rise. The result is looked up in the trig tables under <u>cosine</u> and its corresponding degree is the degree of the fitting.

Find: Degree of bottom elbow.

Solution:

Cosine of degree of rise	=	.707
Cosine of angle of turn	=	X .707
		4949
		4949
		499849 or .5

From the trig tables, the degree that has .5000 for its cosine is 60°. This is the degree of the bottom elbow required for a combination 45° turn and 45" rise in the pipe line as shown above.

The degree of the top elbow is always angle C of triangle ABC in the illustration below. It is the compliment of the degree of rise and therefore is found by subtracting the degree of rise from 90°.

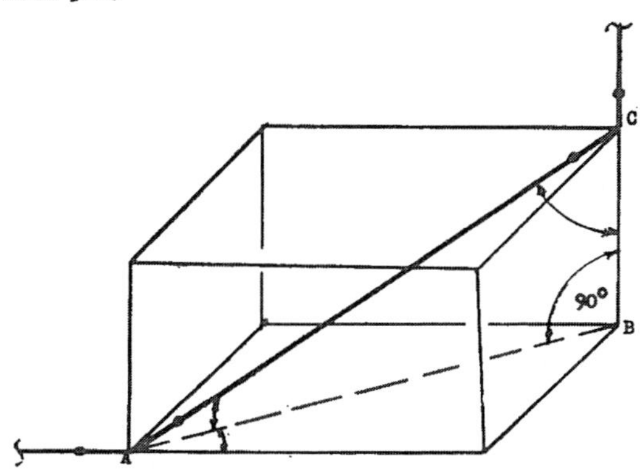

In this problem, the top elbow will be 45°. (90° minus 45° = 45°).

All offsets of this type are solved as described above, regardless of the degree of rise or the angle of turn.

To find the cut length of pipe to go between the two elbows, multiply the cosecant of angle A of triangle ABC times the length of the opposite side BC. This will give you the hypotenuse AC. Then subtract the take up of the 60° bottom elbow and the 45° top elbow from this hypotenuse to obtain the cut length of pipe.

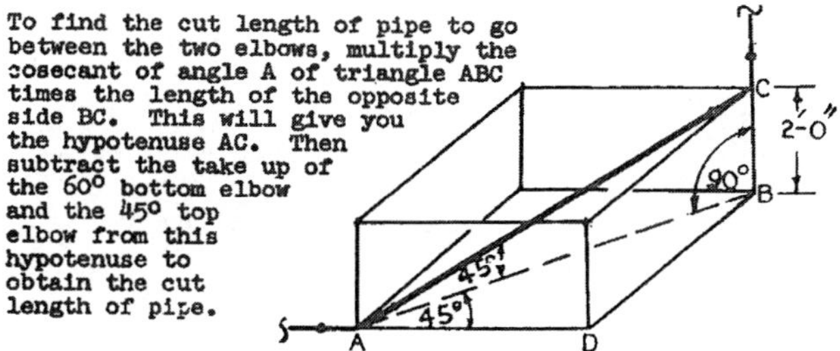

HOW TO FIND DEGREE OF ELBOWS REQUIRED TO MAKE OFFSET THAT RISES AND TURNS AT THE SAME TIME

In the special offset of the design shown below, the degree of the bottom elbow is also found by using the cosine equals cosine formula.

Example problem: Find degree of elbows required to make an offset that has a 30° rise and a turn of 60°.

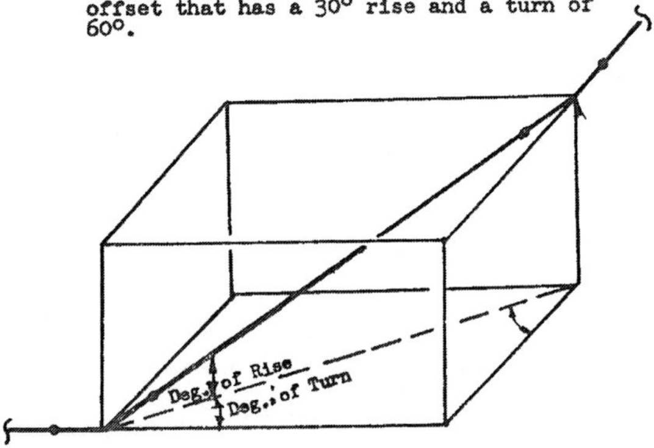

Deg. of Rise
Deg. of Turn

To find degree of bottom elbow: Cosine of 30° (.8660) times cosine of 60° (.5000) equals .4330. From the trig tables the degree that has .4330 for its cosine is 64°-20'. This is the degree of the bottom elbow.

The degree of the top elbow is found by using the same procedure. The only difference is that the top elbow has an angle of turn that is the <u>complement</u> of the degree of turn of the bottom elbow. The degree of rise is always the same for both elbows. Therefore, to find the degree of the top elbow, multiply the cosine of 30° (the compliment of 60°) times the cosine of 30°, the degree of rise. Look up this product under cosine in the trig tables, same as for bottom elbow.

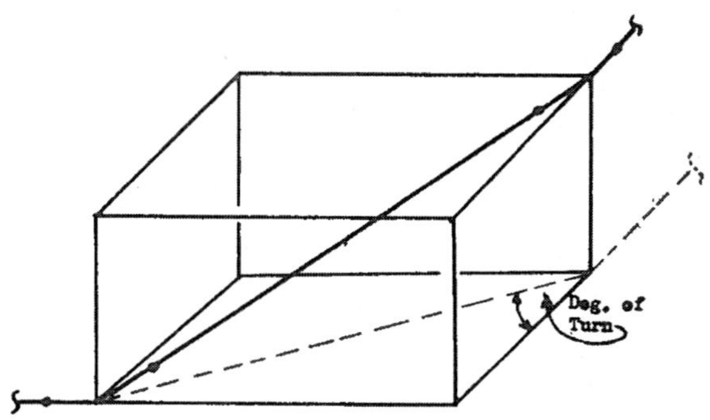

To find degree of top elbow: Cosine of 30° (.866) times cosine of 30° (.866) equals .7499.

From the trig tables, the degree that has .7499 for its cosine is 41°-25'. This is the degree of the top elbow.

FIND ANGLE OF FITTINGS FOR OFFSET LESS THAN RADIUS OF ELBOWS

Example problem:

Find the degree of the 8" elbows that will make a 6" offset when welded together as shown in the illustration below.

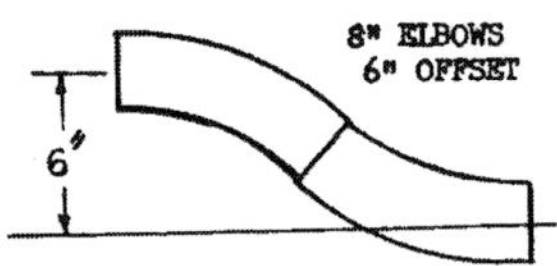

The first step in the solution is to divide the offset into two equal parts by drawing a line through the center of it. See sketch at right.

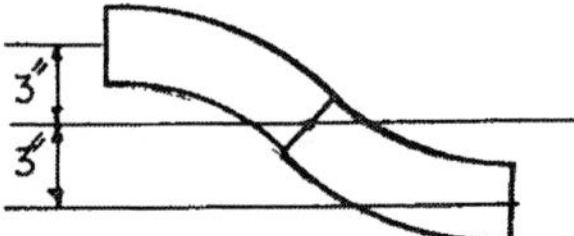

Next draw the radius line, AC, to form right triangle ABC. The object is to find the degree of the elbows, which is angle A of triangle ABC.

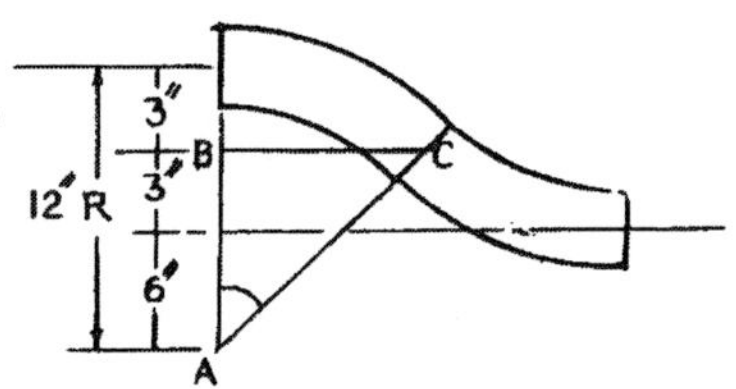

Side AB of triangle ABC is found by subtracting one-half of the offset (3") from the radius of the elbow (12"). It is 9" long and is the adjacent side of angle A of triangle ABC.

Side AC of this triangle is a radius of the elbow (12").
It is also the hypotenuse of triangle ABC.

To find angle A use the formula hypotenuse divided by the adjacent equals secant. In this example problem, AC (12") divided by AB (9") equals 1.333. From the trig tables, the angle that has 1.333 for its secant is 41°-24'. Two 8" elbows, each containing this degree, when welded together will make a 6" offset.

FIND ANGLE OF FITTING FOR OFFSET LESS THAN RADIUS OF ELBOWS
(Use one 90° elbow and one elbow less than 90°)

Given:
7" Offset
6" Welding Elbows

Object: Find degree of the elbow to use with the 90° elbow to make 7" offset.

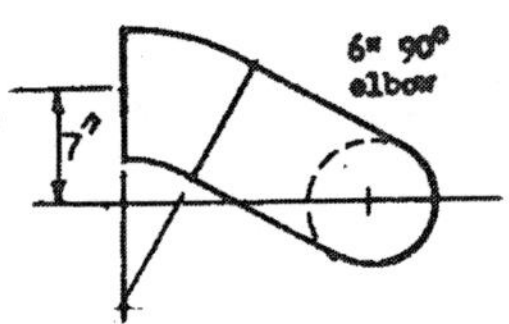

The first step in the solution is to make a rough sketch of the offset and draw in the radius lines of the smaller elbow.

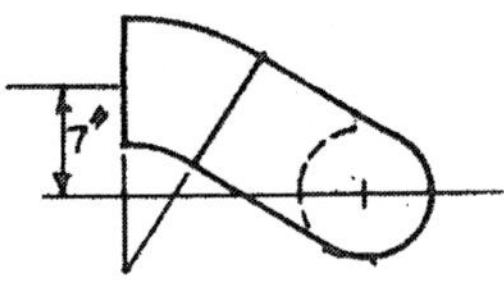

Next draw the radius line of the 90° elbow and the connecting lines AC and BC. Label the triangles thus formed ABC and ADC.

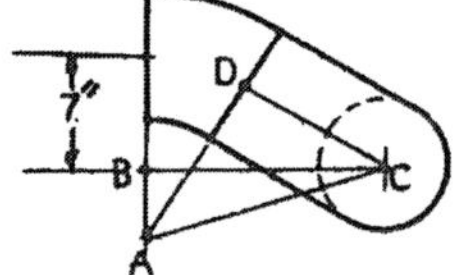

Find angle A of triangle ADC and angle A of triangle ABC. Subtract angle A of triangle ADC from angle A of triangle ABC and the remaining degree will be the degree of the smaller elbow.

Sides AD and DC of triangle ADC are the radius of each of the elbows and are the same length (9"). Therefore, angle A of triangle ADC is 45 degrees.

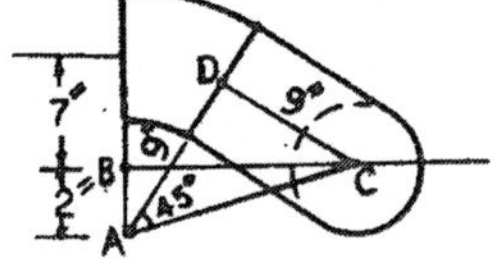

To find side AC, the hypotenuse of triangle ADC, use the formula:

Hypotenuse = Cosecant X Opposite
Hypotenuse = 1.414 X 9
Hypotenuse = 12.72"

To find side AB of triangle ABC, subtract the amount of offset from the radius of the elbow. Side AB is 2".

To find angle A of triangle ABC, use the formula hypotenuse divided by adjacent equals secant:

$$\frac{12.72}{2} = 6.360 \text{ (look up under secant)}$$

The angle that has 6.360 for its secant is 80°-57'. This is angle A of triangle ABC.

To find the degree of the small elbow, subtract angle A of triangle ADC from angle A of triangle ABC.

  80°-57'
- 45°-00'
  35°-57' The degree of the elbow that will make a 7" offset when used with a 90° elbow.

FIND ANGLE OF FITTINGS FOR OFFSET MORE THAN RADIUS OF ELBOWS

Example problem:

Find the degree of the 6" elbows that will make a 12" offset when welded together as shown in the illustration below.

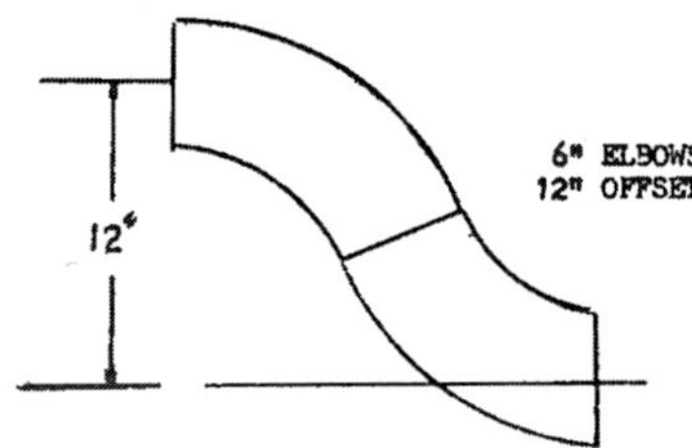

The first step in the solution is to divide the offset into two equal parts by drawing a line thru the center of it. See sketch at right.

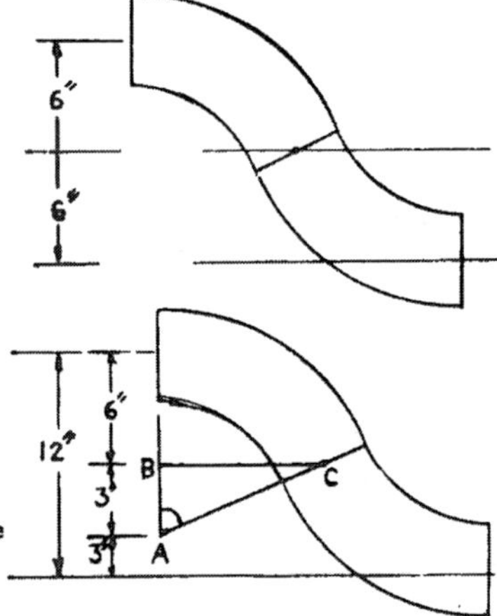

Next, draw the radius line AC to form right triangle ABC. The object is to find angle A, which is the degree of the elbows.

Side AB of triangle ABC is found by subtracting the difference between the amount of offset and the radius of the elbows from one-half of the offset. It is 3" long and is the adjacent side of angle A of triangle ABC.

Side AC of this triangle is a radius of the elbow (9"). It is also the hypotenuse of triangle ABC.

To find angle A, use the formula hypotenuse divided by the adjacent side equals secant. In this example problem, AC (9") divided by side AB (3") equals 3.000. From the trig tables, the angle that has 3.000 for its secant is 70°-32'. Two 6" elbows, each containing this degree, when welded together will make a 12" offset.

FIND ANGLE OF FITTING FOR OFFSET GREATER THAN RADIUS OF ELBOWS
(Use one 90° elbow and one elbow less than 90°)

This is a special offset in which one 90° elbow is rolled up an amount equal to the degree of the smaller elbow. The object is to find the degree of this elbow less than 90° which will make the desired amount of offset when buttwelded against a 90° elbow.

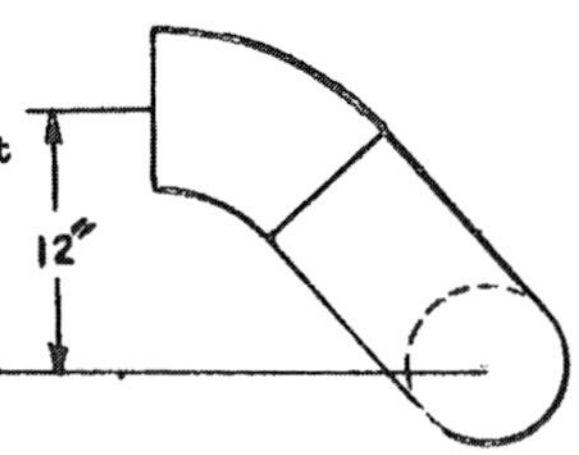

This problem is solved by finding two angles and subtracting their sum from 180 degrees to find a third angle....which is the degree of the fitting. These two angles are acute angles of two right triangles formed by three radius lines, one center line, and two connecting lines all drawn on a sketch of the offset. The angle of the fitting, as always, is the angle at the vertex of radius lines drawn from each end of the smaller elbow.

The first step in solving for the degree of the fitting is to make a rough sketch of the offset and draw in the three radius lines. The angle of the fitting is shown at the vertex of the two radius lines from each end of the smaller elbow. The third line represents the radius of the 90° elbow.

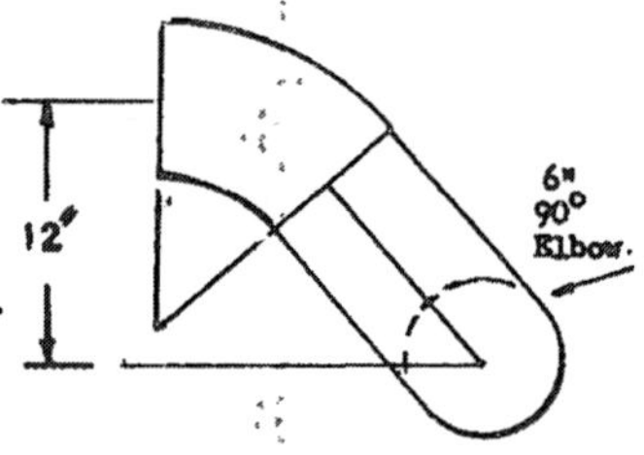

In this example problem, the welding elbows are size 6". Their radius is 9".

Next, draw in the connecting lines AC and AB. Their purpose is to provide sides for the triangles so that angle A of triangle ADC and angle A of triangle ABC can be found. They are the two angles that are subtracted from 180 degrees to obtain the degree of the fitting.

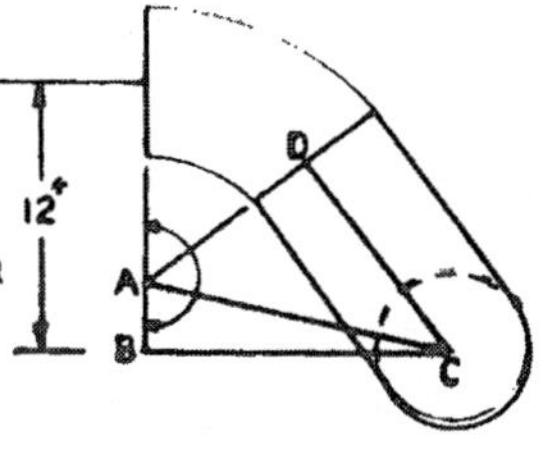

Angle A and angle C of triangle ADC are 45 degree angles. This is because side AD and side DC are each 9" long. (Radius of 6" elbow.)

The length of side AB of triangle ABC is obtained by subtracting the radius of the elbow from the amount of offset. It is 12" minus 9" or 3".

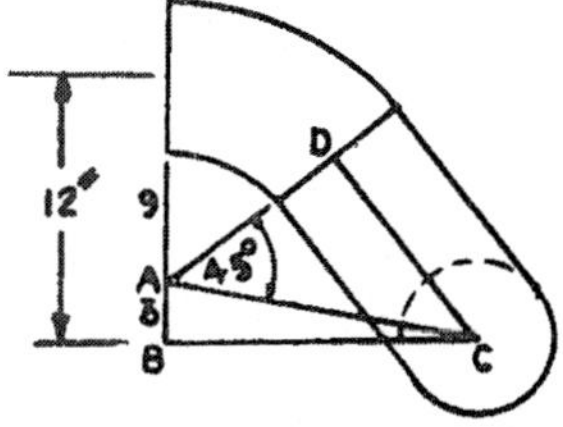

Before finding angle A of triangle ABC, the length of side AC must be determined. In this case it is used as the hypotenuse of triangle ADC because two acute angles and the length of two sides of that triangle are known. This is enough information to enable the length of the third side (AC) to be found. After AC is found it is used as the hypotenuse of triangle ABC. Then angle A of triangle ABC can be determined, because two sides of that right triangle will be known.

To find the length of side AC of triangle ADC use the formula:

Hypotenuse = Cosecant times opposite side
Hypotenuse = 1.414 X 9
Hypotenuse = 12.72" or 12 3/4" (side AC)

Now that the length of side AB and AC are known, solve for angle A of triangle ABC.

Use the formula:

$$\frac{\text{Hypotenuse}}{\text{Adjacent}} = \text{Secant}$$

$$\frac{12.72}{3} = 4.240 \text{ (secant)}$$

In the trig table, look up the angle that has 4.240 for its secant. It is 76°-22'.

$$\begin{array}{r} 76^\circ\text{-}22' \\ +\ 45^\circ\text{-}00' \\ \hline 121^\circ\text{-}22' \end{array}$$ sum of both angle A's

$$\begin{array}{r} 179^\circ\text{-}60' \\ -121^\circ\text{-}22' \\ \hline 58^\circ\text{-}38' \end{array}$$ the degree of the 6" elbow that will make 12" offset when used with 90° elbow.

HOW TO FIND "TRAVEL AHEAD" NECESSARY TO MAKE EQUAL SPREAD OFFSETS

Each of the center to face of flange lengths of the dimensioned ends of the 45° offsets in the sketch below have different measurements. The amount that each measurement is longer than the other is called "travel ahead". This increase in length from one pipe to the next one is necessary so that the lines will remain the same distance apart after they are turned to make the offsets. Because the lines are the same distance apart, the travel ahead will be the same from one offset to the next one. This travel ahead is found by solving for side BC in triangle ABC (below).

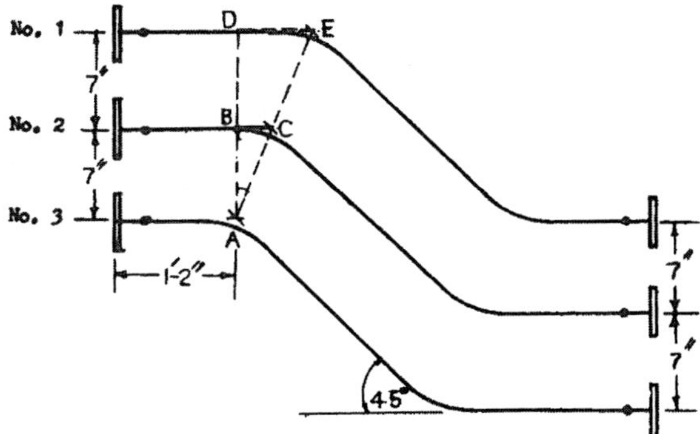

Side BC is the opposite side of angle A in triangle ABC. Angle A is one-half of the degree of turn or 22 1/2 degrees. Side AB is the adjacent side of angle A.

Length of side opposite (BC) = Side adjacent (AB) X tangent
BC = 7 x .4142
BC = 2.899" or 2 15/16"

2 15/16" + 1'-2" = 1'-4 15/16" Center to face of flange of offset No. 2.

1'-4 15/16" + 2 15/16" = 1'-7 7/8" Center to face of offset No. 1.

Side DE of triangle ADE is twice as long as length of BC of triangle ABC. If there were four offsets, the travel ahead for the fourth one would be three times the length of BC. This holds true whenever the offsets are the same distance apart.

HOW TO FIND "TRAVEL AHEAD" NECESSARY TO MAKE EQUAL SPREAD OFFSETS

The sketch below is an equal spread offset that has different center to center spacing. To find the center to face dimension of offset number 2, solve for side BC in right triangle ABC and add this length to the center to face dimension of offset number 3. To obtain the center to face dimension of offset number 1, find side DE of triangle ADE and add this length to the center to face dimension of offset number 3.

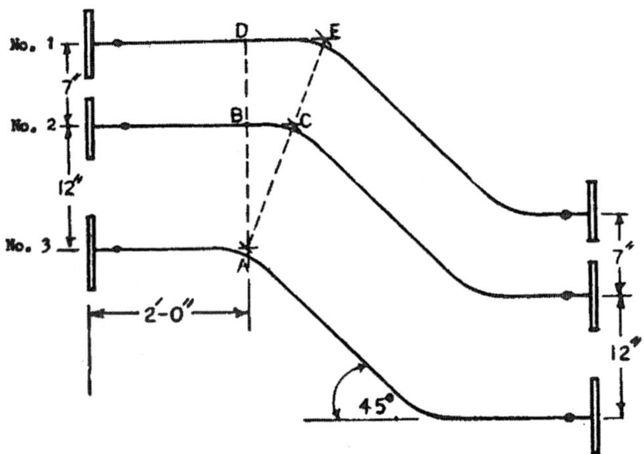

Side BC is the opposite side of angle A in triangle ABC. Angle A is one-half of the degree of turn or 22 1/2 degrees. Side AB is the adjacent side of angle A.

Length of side opposite (BC) = Side adjacent (AB) times tangent of angle A

BC = 12" X .4142

BC = 4.97" or 5"

5" plus 2'-0" = 2'-5" Center to face of flange of offset number 2

Side DE is the opposite side of angle A in triangle ADE. Angle A is one-half the degree of turn or 22 1/2 degrees. Side AD is the adjacent side of angle A and is 19" long.

Length of side opposite (DE) = Side adjacent (AD) times tangent of angle A

DE = 19" X .4142

DE = 7.869" or 7 7/8"

7 7/8" plus 2'-0" = 2'-7/8" Center to face of flange of offset number 1

HOW TO SOLVE 45 DEGREE UNEQUAL SPREAD OFFSETS

The object is to find the horizontal center to center distance between the center of the top elbow of offset No. 1 and the top elbow of offset No. 2 as shown below. This is a certain dimension that will cause the pipe lines to be 10" apart after they are turned on 45 degrees.

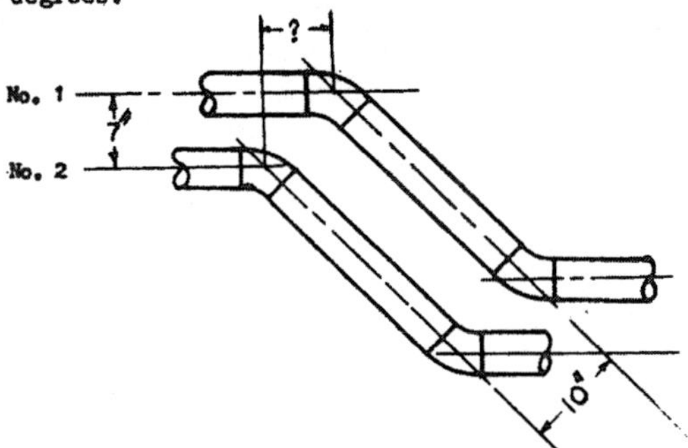

To find this difference in centers, it is necessary to construct and solve two right triangles. The first step is to draw line AC from the center of the top elbow of offset No. 2 to where it intersects the center line of offset No. 1 at point C. See sketch below.

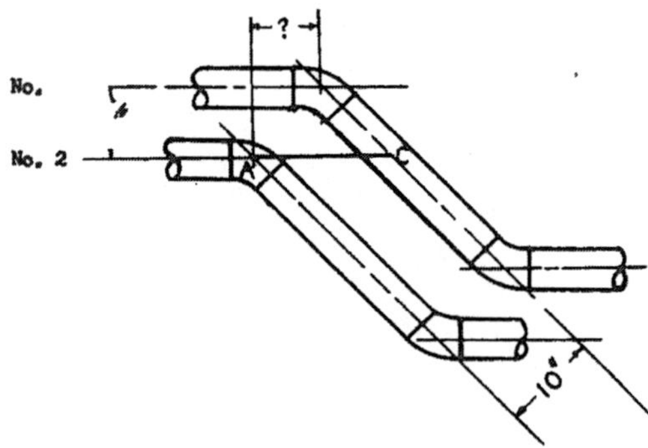

The next step is to draw line CB perpendicular to the center lines of the pipe between the 45 degree elbows. Then draw a line between point A and point B along the center line to form triangle ABC.

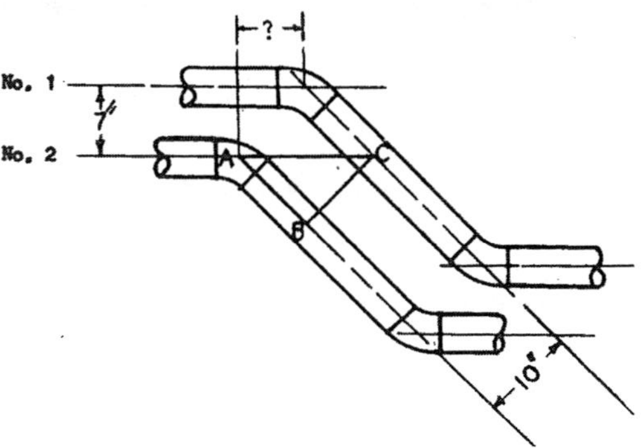

Next, draw perpendicular line DE from the center of the 45 degree elbow to where it intersects line AC at point E. Draw line DC along the center line to form right triangles DEC (below).

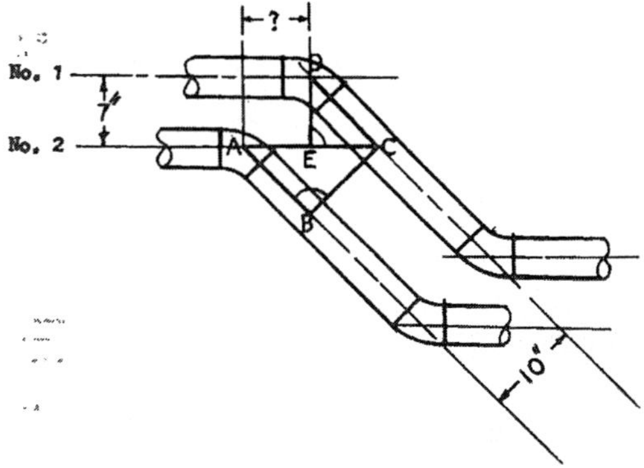

The object of the problem is to find the length of line AE, which is the difference in centers between the elbows.

First find side AC of triangle ABC. Next, find side EC of triangle DEC and then subtract EC from AC to find the distance AE.

Solution: Angle D of triangle DEC and angle A of triangle ABC are 45 degree angles. This is because the angles of turn are 45 degrees.

In triangle ABC:

Hypotenuse (AC) = Side opposite X cosecant of angle A.
AC = 10" X 1.414
AC = 14.14"

In triangle DEC:

Opposite side (EC) = Side adjacent X tangent
EC = 7" X 1.000
EC = 7"

AC (14.14" minus EC (7") = 7.14" (A E)

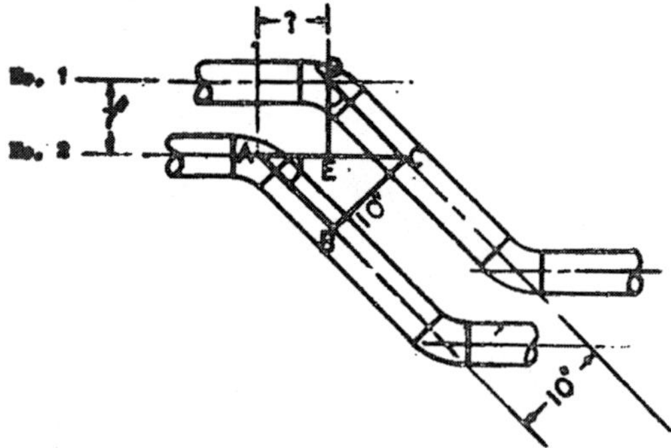

When the center of the 45 degree elbow of offset No. 1 is 7.14" ahead of the 45 degree elbow of offset No. 2, the distance BC between the run pipes will be 10".

HOW TO SOLVE 45 DEGREE UNEQUAL SPREAD OFFSETS

The object is to find the horizontal difference in centers between the two 45 degree elbows. The length of the pipe required to make the two flanges flush with each other is the same as the difference in centers.

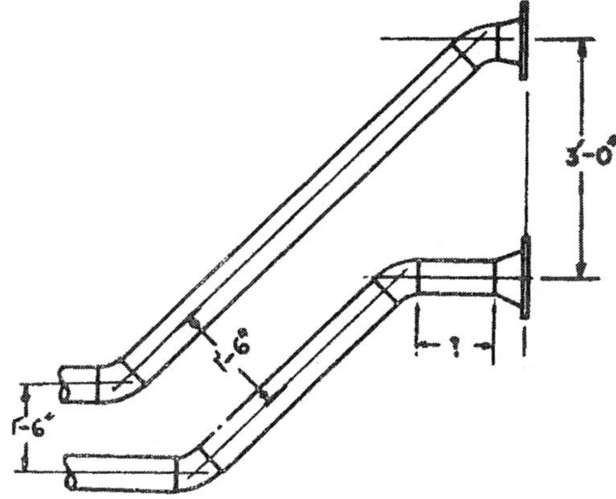

To find this difference in centers, it is necessary to construct and solve two right triangles, shown on the next page. The first step is to drop a perpendicular from the center of the top 45 degree elbow to where it meets the center line of the other offset at point B. Then extend the center line from point B to where it intersects the center of the pipe in the top offset at point A. This forms triangle ABC.

From point A, draw a line perpendicular to the center line of the pipe between the elbows to point D. Draw line DE to form 45 degree right triangle ADE.

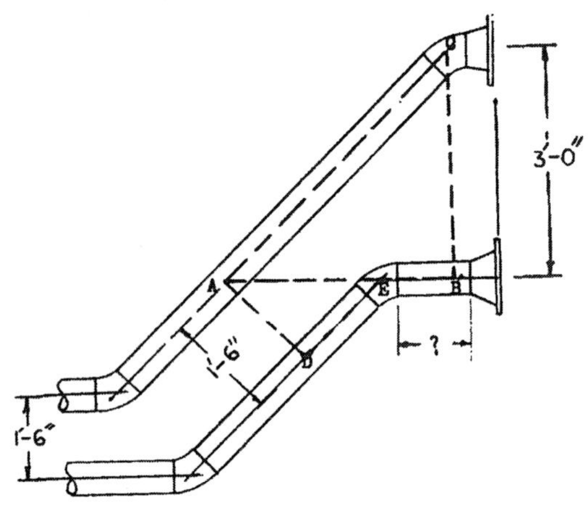

Line EB is the difference in centers between the two elbows. To obtain its length, find side AB of triangle ABC and side AE of triangle ADE. Subtrace AE from AB to obtain the length of EB.

Side AB is 3'-0" long because the legs of 45 degree right triangles are always the same length. Side AE is 1.414 times 18" or 25.45". (Hypotenuse = Secant times adjacent side.)

AB 36.00"
Minus AE 25.45"
10.55" or 10 9/16" Length of EB

The difference in centers is 10 9/16" and is also the length of the pipe between the 45 degree elbow and the flange.

HOW TO SOLVE SPECIAL TANK OFFSET

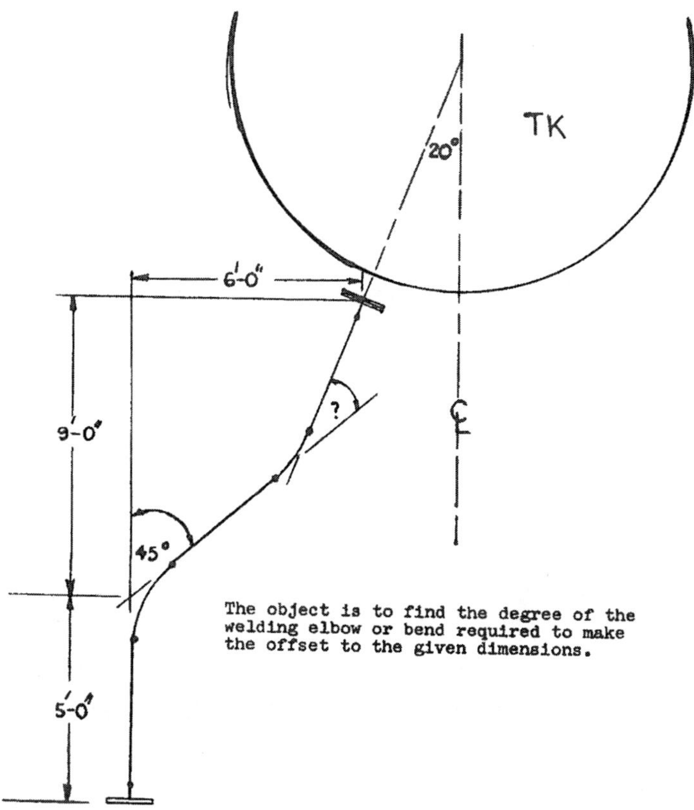

The object is to find the degree of the welding elbow or bend required to make the offset to the given dimensions.

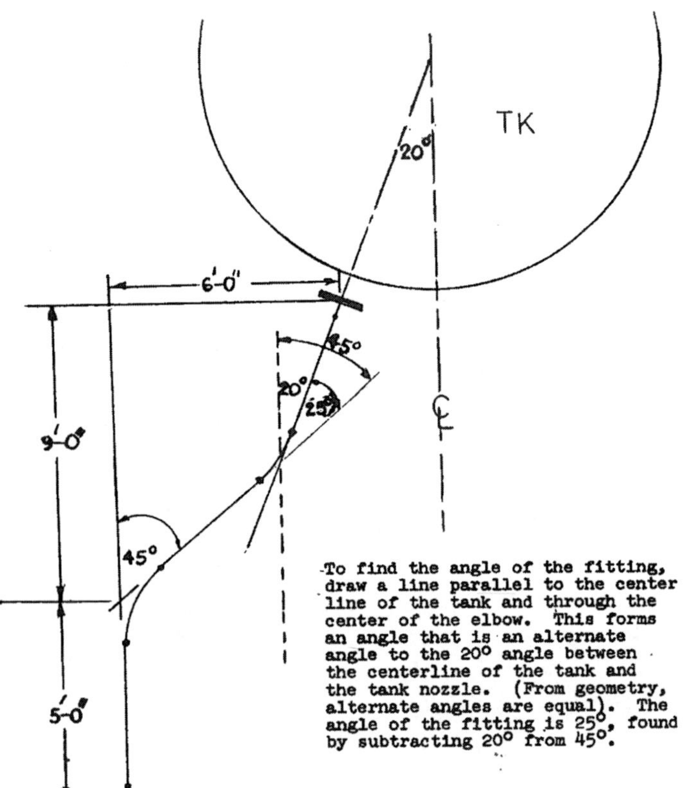

To find the angle of the fitting, draw a line parallel to the center line of the tank and through the center of the elbow. This forms an angle that is an alternate angle to the 20° angle between the centerline of the tank and the tank nozzle. (From geometry, alternate angles are equal). The angle of the fitting is 25°, found by subtracting 20° from 45°.

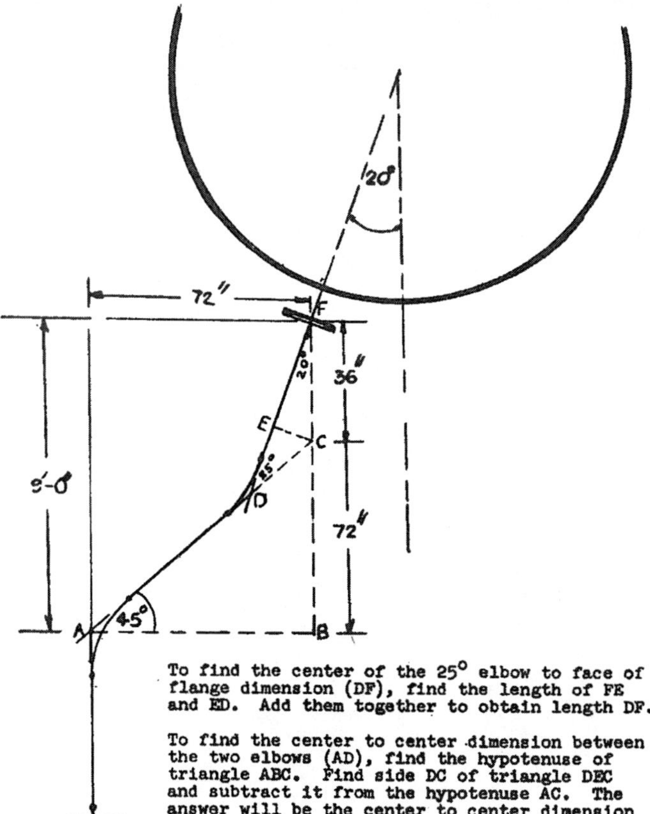

To find the center of the 25° elbow to face of flange dimension (DF), find the length of FE and ED. Add them together to obtain length DF.

To find the center to center dimension between the two elbows (AD), find the hypotenuse of triangle ABC. Find side DC of triangle DEC and subtract it from the hypotenuse AC. The answer will be the center to center dimension between the two elbows (AD).

HOW TO FIND THE DIMENSION FROM THE CENTER OF THE ELBOW TO THE FACE OF THE FLANGE ON THE TANK.

To form the three right triangles shown on the preceding page, draw line FB parallel to the center line of the tank. Extend line AD until it intersects line FB at point C. Draw line EC perpendicular to DF.

Side BC of Triangle ABC is 72" because AB is 72" and angle A and C are both 45°. Therefore, side FC of triangle FEC is 36" long. (9'-0" minus 6'-0").

Multiply side FC of triangle FEC times the cosine of 20° and their product is 33.83"-----which is the length of FE.

Multiply side FC times the sine of 20° to obtain side EC of triangle FEC. Side EC is 12.31".

In triangle DEC, multiply the cotangent of 25° times the opposite side EC (12.31") to obtain side DE. It is 26.40".

FE	33.83
+ DE	26.40
	50.23" or 5'-0 1/4" length of DF

To obtain the distance between the two elbows:

In triangle ABC multiply the cosecant of 45° times the opposite side BC to obtain the hypotenuse AC. It is 101.8".

To find side DC of triangle DEC multiply the cosecant of 25° times opposite side EC (12.31"). It is 29.02". Subtract DC from AC to obtain side AD.

AC	101.8"
DC	29.02"
	72.78" or 6'-0 3/4" length of AD center to center between the two elbows.

FIND LENGTH OF SEGMENTS OF FOUR TURN 90° BEND WHEN RADIUS IS NOT GIVEN

Notice that the length of the radius of the 90° bend is not given. The only dimension that is known is distance from the center of one bend to the center of the pipe after the 90° turn is made. (See sketch at right). Before the center to center length of each segment can be found, the radius of turn must be determined.

This is accomplished by solving three right triangles. The length of the segments can then be found in the usual manner.

The first step is to draw a line thru the center of the outside bends (shown dotted). These lines are perpendicular to each other. Draw the hypotenuse line to form a right triangle, the legs of which are 120" long, the only given dimension. Find this hypotenuse by multiplying 120 times the cosecant of 45°. It is 169.68". The purpose in finding this length is so that line AB can be found by dividing it into two equal parts.

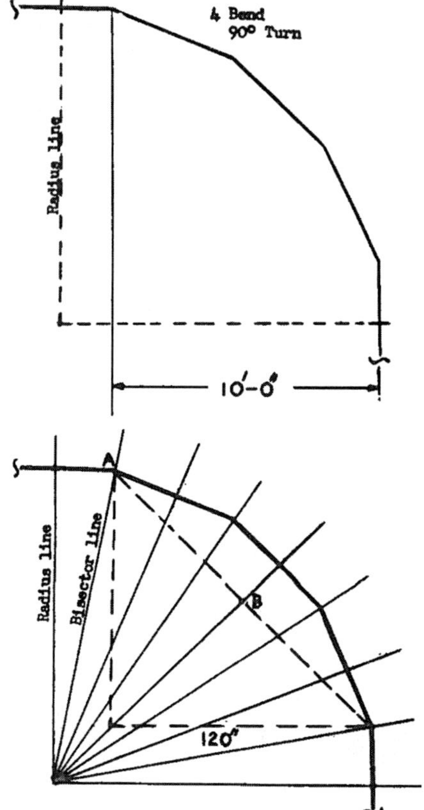

Line AB is 84.84"-- one-half of 169.68". Next draw right triangle ABC by using bisector line AC and part of a radius line as CB.

Each of the three small angles that make up angle C are 11°-15'. (One half of the angle of turn of each bend.) Therefore, angle C is 33°-45'.

Find side AC by multiplying the cosecant of 33°-45' times 84.84", the opposite side of angle C of triangle ABC. 7"

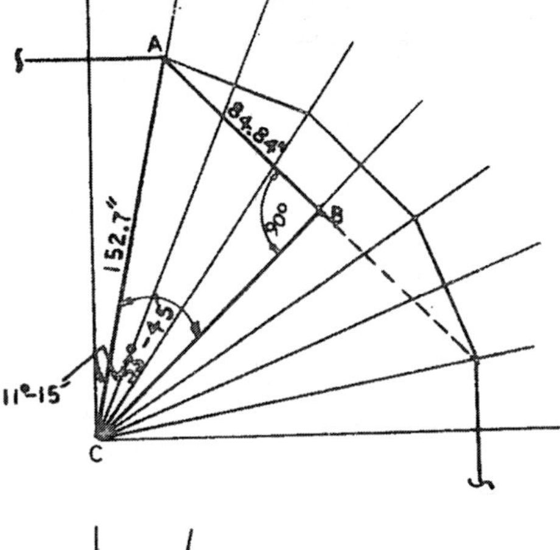

The sides of triangle ADC are made up of radius line CD, bisector CA and the opposite side of angle C, DA. To find the length of the radius line, multiply the cosine of angle C times the hypotenuse AC. (Adjacent side equals cosine times hypotenuse.) The radius of the bend, side CD of the triangle is 149.76".

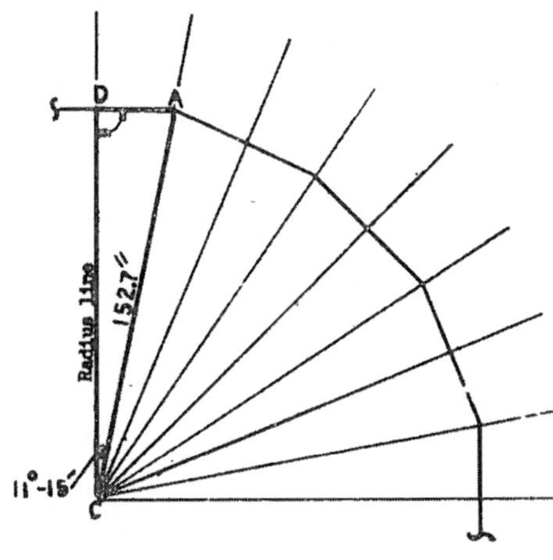

To find the length of the segments of the bend, find side DA of triangle ADC and multiply by 2.

Each segment is the opposite side of two right triangles identical to triangle ADC. To find the length of side DA, multiply the tangent of angle C times the adjacent side, CD.

Tangent of 11°-15' (.19891) times 149.76 times 2 equals 59.58" or 4'-11 5/8". This is the center to center length of each segment.

HOW TO FIND CUT BACK LENGTHS FOR 60° "Y" TYPE INTERSECTION

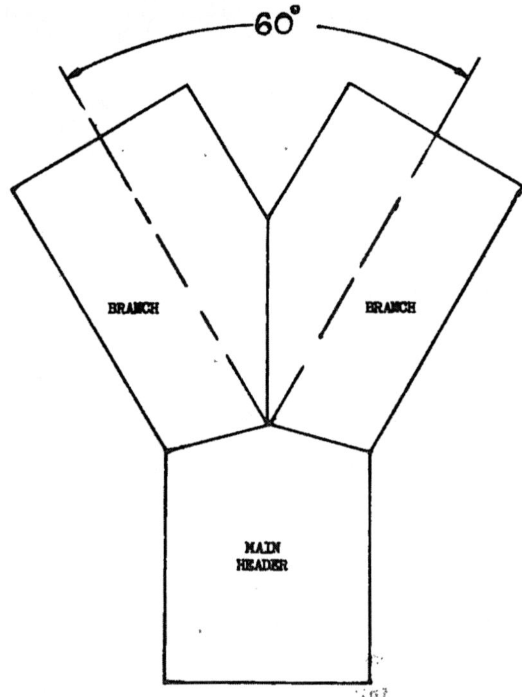

As seen above, the "Y" type intersection consists of three pieces. The main piece has two 15° miter cuts on it and the two branches each have a 60° and a 15° miter cut that fit together to form the intersection.

To lay out the main header:

Divide the circumference of the pipe into 16 equal parts and draw 16 ordinate lines. Then lay off two 15° miter lines on one-half of the pipe as shown below.

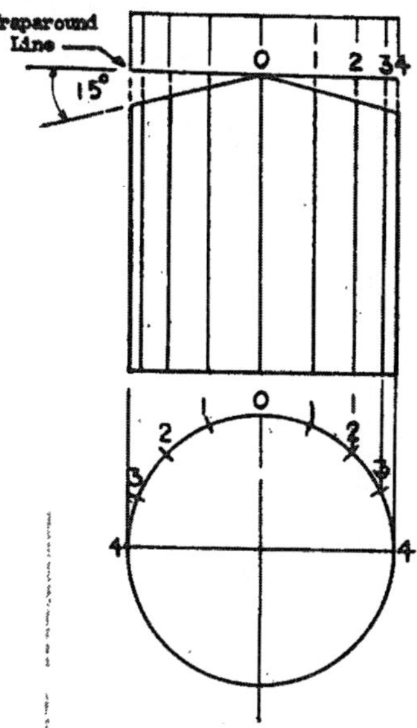

To find cut back lengths for 15° miter:

Line No. 4----Multiply tangent of 15° times 1/2 the pipe I.D.

Line No. 3----Multiply .923 times No. 4 cut back length obtained above.

Line No. 2----Multiply .707 times No. 4 cut back length.

Line No. 1----Multiply .382 times No. 4 cut back length.

How to lay out the two branch pipes:

Lay out a 60° miter line on one-half of the pipe and a 15° miter line on the other half as shown in the sketch below. Use the same formula for finding cut back lengths as was used on the main header.

This layout is for a straight in, radial cut. If hacksaw type burn is to be made, use one-half the OUTSIDE DIAMETER of the pipe when figuring cut back lengths.

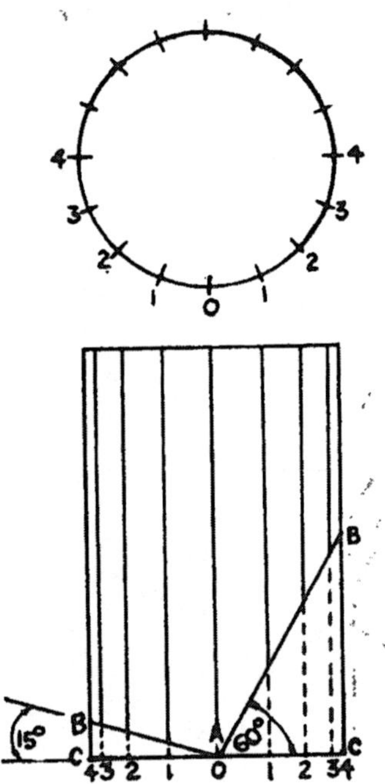

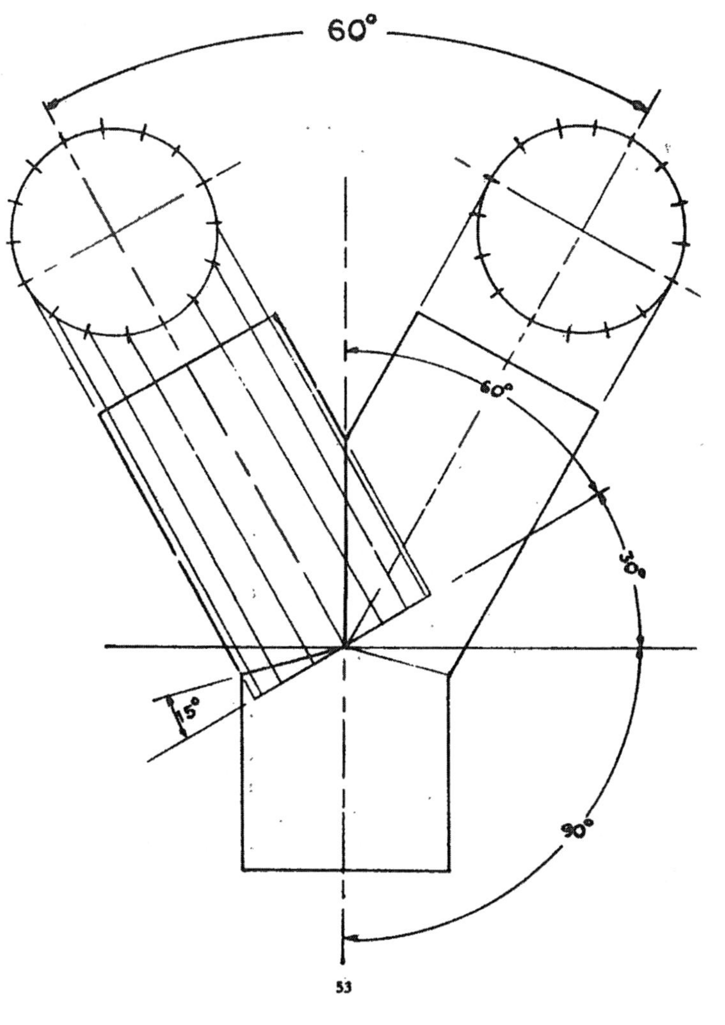
60°
60°
30°
15°
90°

SKETCH SHOWING LAYOUT OF 45 DEGREE "Y" TYPE INTERSECTION

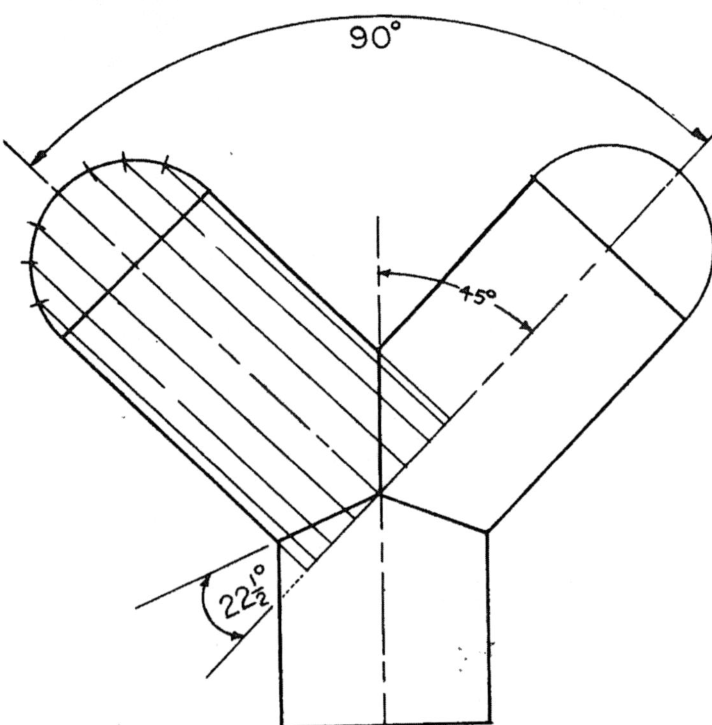

The 45° type intersection is fabricated in the same manner as the 60° intersection. The only difference is that the miter cuts are 45° and 22 1/2°.

HOW TO FIND ORDINATE LENGTHS FOR NINETY DEGREE PIPE INTERSECTIONS

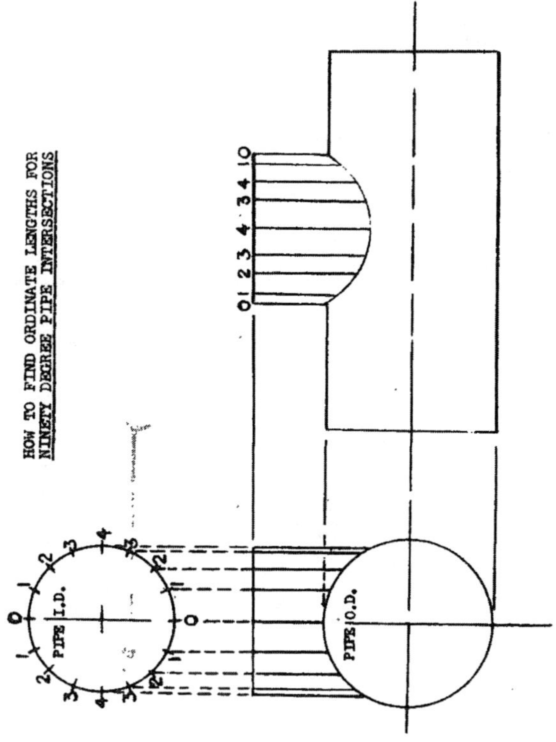

HOW TO FIND ORDINATE LENGTHS FOR 90° PIPE INTERSECTIONS

Ordinate lengths are drop down distances that are measured from a wraparound line on the riser pipe. They are marked off on ordinate lines that are equally spaced around the circumference of the pipe. When the ends of their different lengths are connected together with a soapstone marking pencil, a curving line is formed that will fit the contour of the outside surface of the header pipe. Depending upon the size of the pipe, eight or sixteen ordinate lines and ordinate lengths are used.

When sixteen ordinate lines are used, they are numbered 0 thru 4 with the two lines numbered 0 being the throat of the saddle. The two lines numbered 4 are the side centers...the long points of the saddle. Since these ordinate lines are marked off in four quarters of the pipe, this makes a total of 16 lines around the pipe.

When eight ordinate lines are used, they are numbered 0, 1, and 2, with number 2 line being the long points of the saddle.

Ordinate number 0 does not have length because there is no drop down at all. This is because it is in the throat of the saddle and touches the header on top center. As stated above, number 4 ordinate is the longest because it is on the side center of the saddle.

Most of the intersections made today are of the "saddle on" type. This means that the inside diameter of the riser is fitted against the outside surface of the header pipe. For this reason, the <u>inside radius</u> of the riser and the <u>outside radius</u> of the header are used when figuring these ordinate lengths. To find these lengths involves solving two right triangles. The hypotenuse of one is the inside radius of the riser and the hypotenuse of the other is the outside radius of the header

To determine the length of ordinate No. 1, find side FK in triangle EKF. Then subtract FK from one-half the O.D. of the header and the remainder will be ordinate No. 1.

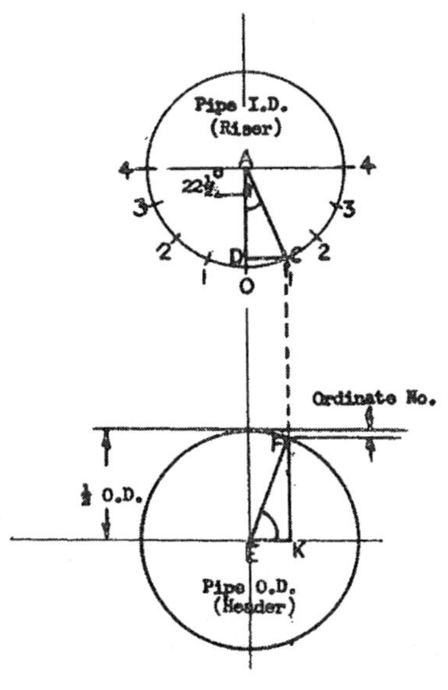

Side DC in triangle ADC must be known before FK can be found. Side AC is the radius of the I.D. of the riser pipe. Angle A is 22½ degrees because the I.D. circle is divided into 16 equal parts by the radius lines. Find opposite side DC by multiplying the sine of 22½ degrees by the hypotenuse AC.

Side DC is projected down into the O.D. circle to become side EK in triangle EKF. Side EF is the radius of the O.D. of the header. Divide hypotenuse EF by adjacent side EK to find angle E. (Look up under secant in trig tables.

To find opposite side FK, multiply the sine of angle E times the hypotenuse EF. Subtract FK from ½ of the outside diameter of the header pipe to obtain the length of No. 1 ordinate.

Ordinate line number 2 and 3 are found in exactly the same manner as ordinate number 1. The only difference is that angle A of triangle ADC will be 45 degrees for number 2 line and 67½ degrees for number 3 line.

Find the length of ordinate number 2 by using the same procedure used for ordinate number 1. Angle A of triangle ADC will be 45 degrees because it is made up of two 22½ degree divisions of the I.D. circle. The steps are as follows:

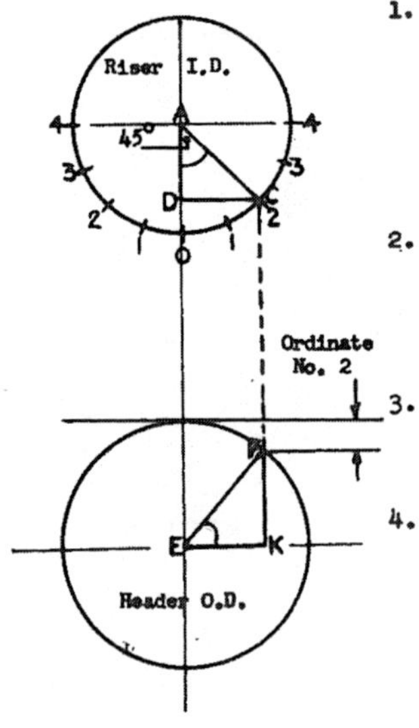

1. Find side DC in triangle ADC by multiplying the sine of 45 degrees times one-half the inside diameter of the riser. Project side DC down into the O.D. circle to become side EK in triangle EKF.

2. Find angle E of triangle EKF by dividing side EF, which is one-half the outside diameter of the header by side EK.

3. Find side FK by multiplying the sine of angle E times the hypotenuse EF.

4. Subtract side FK from one-half the outside diameter of the header and this difference will be the length of ordinate number 2.

The length of ordinate number 3 is found by using the same method used to find ordinate number 1 and 2. Angle A of triangle ADC will be 67½ degrees because it is made up of three 22½ degree divisions of the I.D. circle.

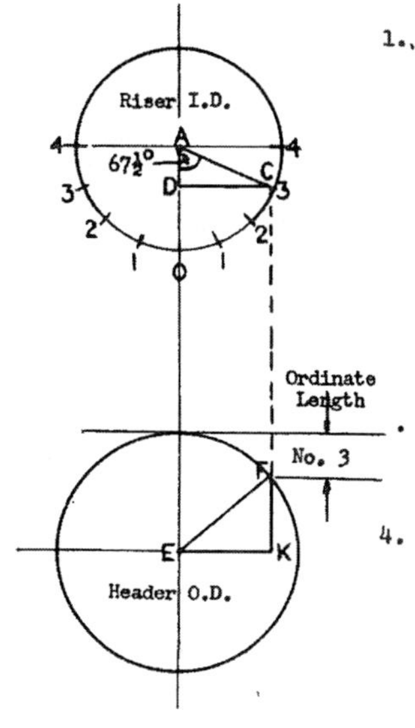

1. Find side DC in triangle ADC by multiplying the sine of 67½ degrees times one-half the inside diameter of the riser. Project side DC into the O.D. circle to become side EK in triangle EKF.

   Find angle E of triangle EKF by dividing side EF, which is one-half the outside diameter of the header, by side EK. Look up under secant.

. Find side FK by multiplying the sine of angle E times the hypotenuse EF.

4. Subtract side FK from one-half the outside diameter of the header and this difference will be the length of ordinate number 3.

In finding ordinate number 4, it is necessary to solve only one right triangle. This is because side EK in triangle EKF is projected down from the I.D. circle of which it is the radius. The steps in solving for ordinate number 4 are as follows:

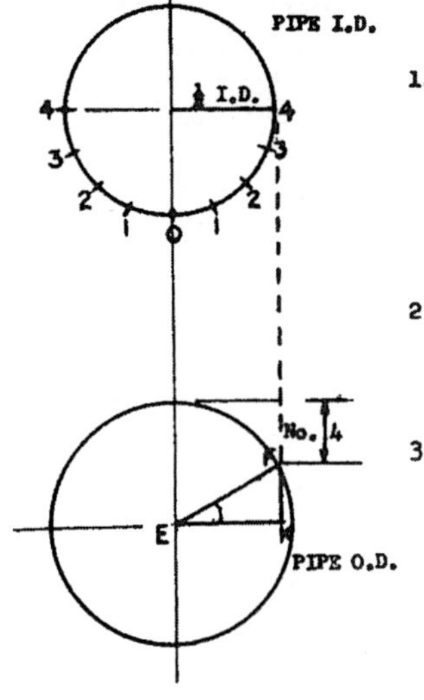

1. Find angle E of triangle EKF by dividing side EF, which is one-half the outside diameter of the header by side EK, which is one-half the inside diameter of the riser.

2. Find side FK by multiplying the sine of angle E times the hypotenuse EF.

3. Subtract side FK from one-half the outside diameter of the header and this difference will be the length of ordinate number 4.

HOW TO FIND EIGHT ORDINATE LENGTHS FOR 45° PIPE INTERSECTIONS

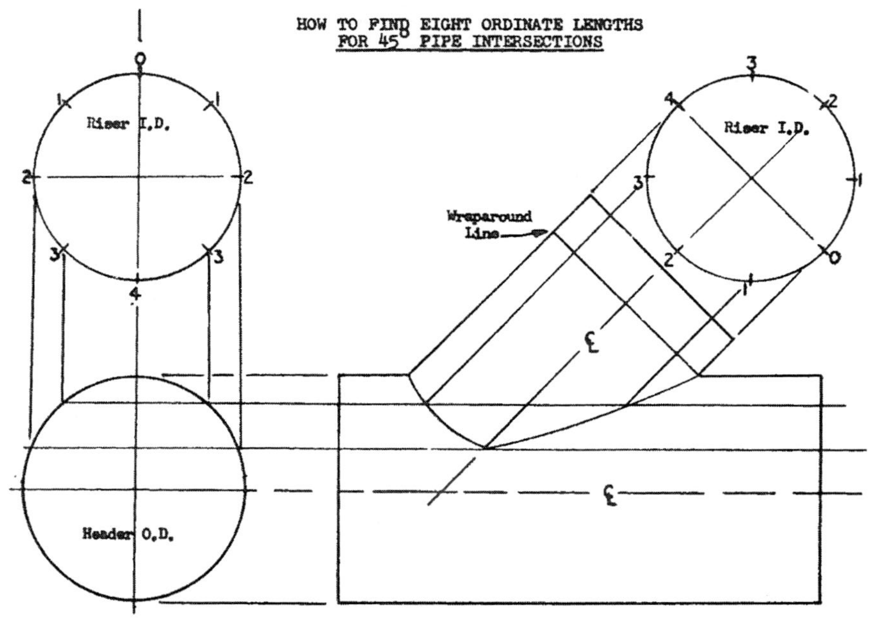

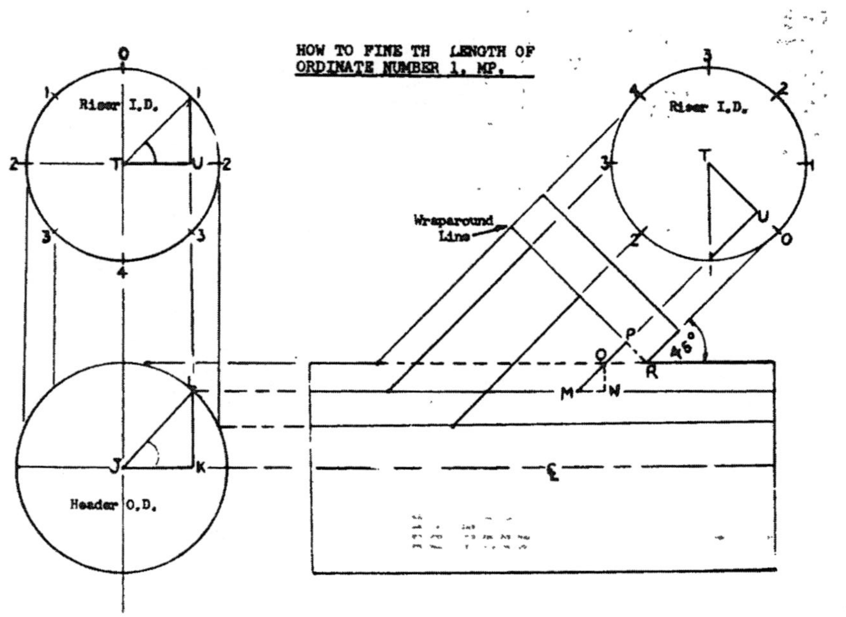

HOW TO FINE TH LENGTH OF
ORDINATE NUMBER 1, MP.
Riser I.D.
Riser I.D.
Header O.D.
Wraparound Line
45°

Ordinate number 1 is made up of side OM of triangle MNO and side PO of triangle OPR. Its length is found by finding these two sides and adding them together.

Side ON, which must be known before OM can be found, is the same length as the number 1 90° ordinate length. It is necessary to find the 90° ordinates before proceeding to find the ordinate lengths of the 45° intersection.

How to find the length of ordinate number 1, MP:

1. Find the 90° ordinate length by finding side LK in triangle JKL and subtracting it from one-half the O.D. of the header. Project this length over to the side view where it becomes side ON of triangle MNO.

2. Find side OM of triangle MNO by multiplying the cosecant of angle M (45°) times ON.

3. Find side PO of triangle OPR. PR is the same length as U0 and U2 in the I.D. circles. PO is the same length as PR because triangle OPR is a 45° right triangle.

4. Add length OM to the length of PO to obtain ordinate number 1, MP.

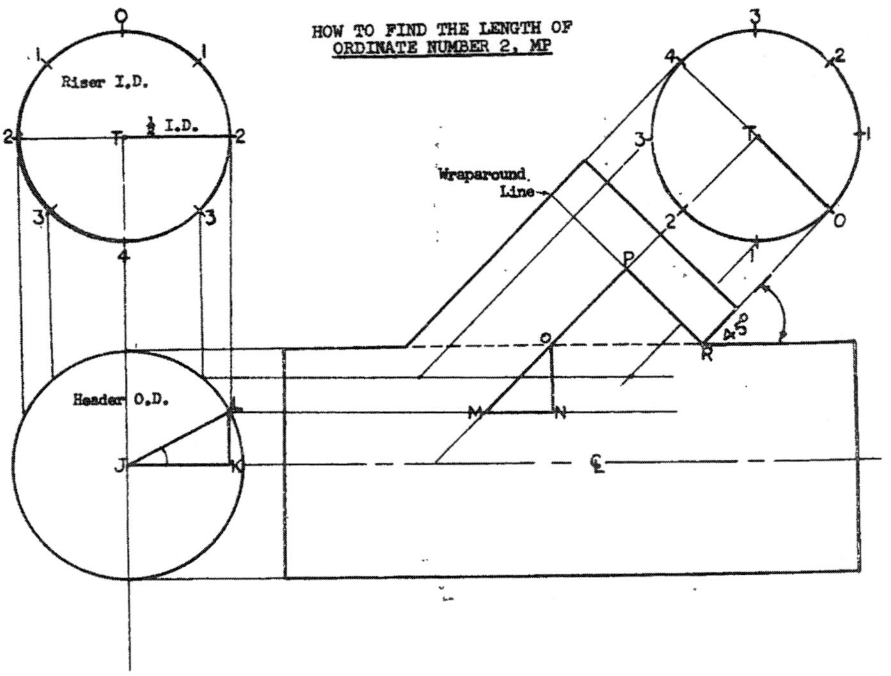
HOW TO FIND THE LENGTH OF
ORDINATE NUMBER 2, MP
Riser I.D.
½ I.D.
Header O.D.
Wraparound
Line
45°

How to find the length of ordinate number 2, MP:

1. Find the 90° ordinate length using the same procedure as for ordinate number 1. Project this length over to the side view where it becomes side ON of triangle MNO.

2. Find side OM of triangle MNO by multiplying the co-secant of angle M (45°) times ON.

3. Find side PO of triangle OPR. PR is the same length as T2 and TO (one-half the riser I.D.). PO is the same length as PR because triangle OPR is a 45° right triangle.

4. Add length OM to the length of PO to obtain ordinate number 2, MP.

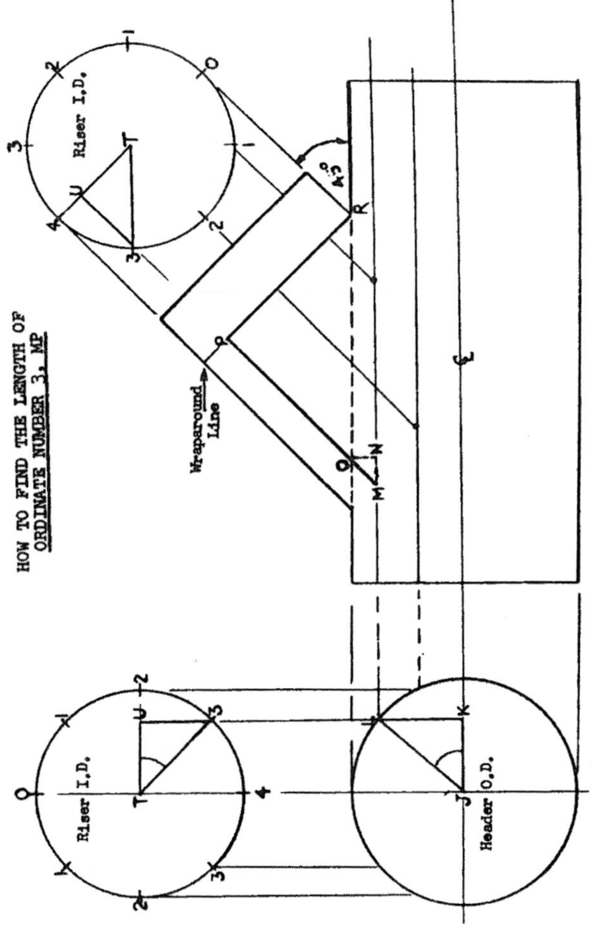
HOW TO FIND THE LENGTH OF
ORDINATE NUMBER 3, MP
Riser I.D.
Riser I.D.
Header O.D.
Wraparound Line
45°

How to find the length of ordinate number 3, MP:

1. Side ON of triangle MNO is the same length as side ON previously found for ordinate number 1. This is because there are two number 3 ordinates for a 90° intersection and they are the same length. Project this length over to the side view where it becomes side ON of triangle MNO.

2. Find side OM of triangle MNO by multiplying the cosecant of angle M (45°) times ON.

3. Find side PO of triangle OPR. Side PR of triangle OPR is found by subtracting U4 (which is the same length as U2) from the inside diameter of the riser pipe. PR and PO are the same length because triangle OPR is a 45° right triangle.

4. Add length OM to the length of PO to obtain ordinate number 3, MP.

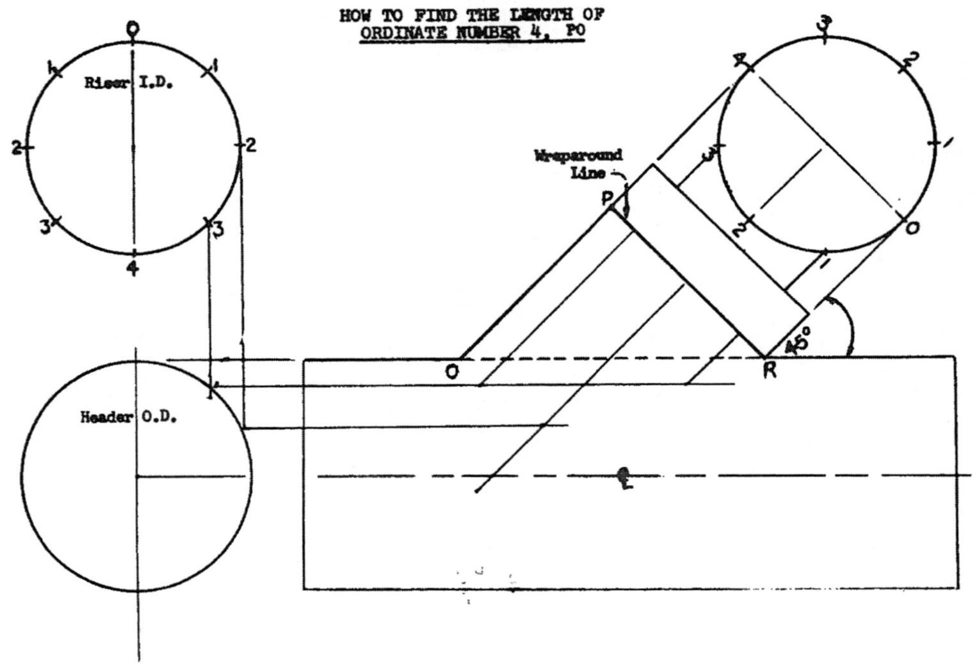
HOW TO FIND THE LENGTH OF
ORDINATE NUMBER 4, PO
Riser I.D.
Header O.D.
Wraparound Line
P
O
R
45°

How to find the length of ordinate number 4, PO:

1. To obtain the length of ordinate number 4, it is only necessary to solve one right triangle. It is triangle OPR shown in the sketch on the preceding page.

2. Side PR in triangle OPR is the same as the inside diameter of the riser pipe. Side PO is the same length as PR because OPR is a 45° right triangle. Number 4 ordinate (or number 8 when using 16 ordinates) will always be the same dimension as the inside diameter of the riser pipe. It is not necessary to consider the right triangle OPR to determine this ordinate length.

When using 16 ordinate lines the same procedure is used for finding the ordinate lengths. The only difference is that the acute angles in the triangles in the I.D. circles will be 67½ degrees and 22½ degrees instead of the 45 degree angles formed when the circle is divided into eight parts.

This entire procedure is not as lengthy as it first seems because the same trigonometry formulas can be used over and over and the process becomes routine. Also, one dimension, the length of line U2, is used a number of times in the solution.

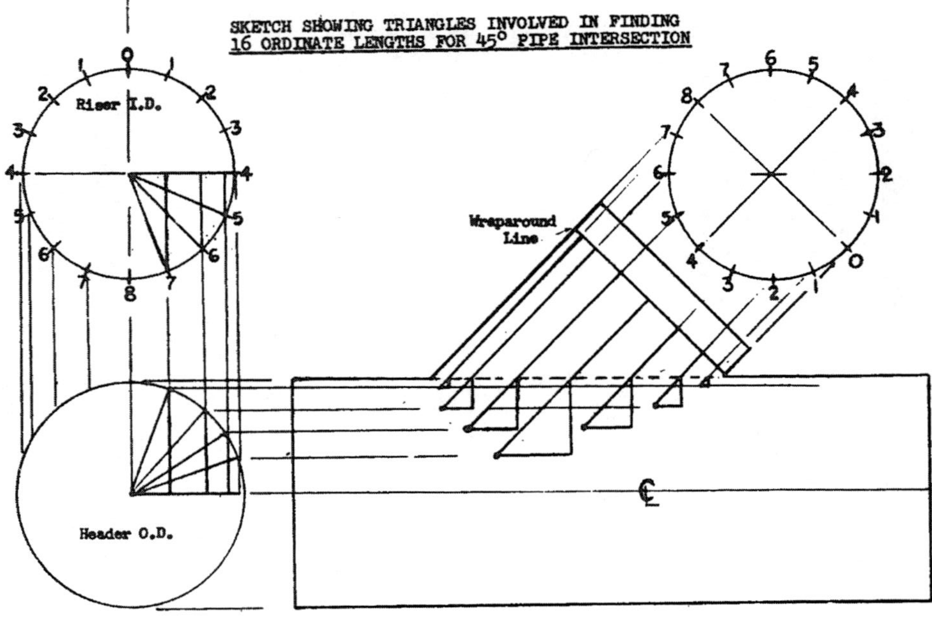

SKETCH SHOWING TRIANGLES INVOLVED IN FINDING 16 ORDINATE LENGTHS FOR 45° PIPE INTERSECTION

HOW TO FIND EIGHT ORDINATE LENGTHS FOR OFF-CENTER INTERSECTIONS

Some jobs call for 90° pipe risers to be installed off of the center of the header, but not tangent to the header pipe wall (eccentric).

The ordinate lengths for this type of intersection are calculated by using the same method used to find 90° concentric ordinate lengths. However, the difference in centers between the riser and header must be taken into consideration when making these calculations. In the sketch below, the riser pipe is ½" off of the center of the header.

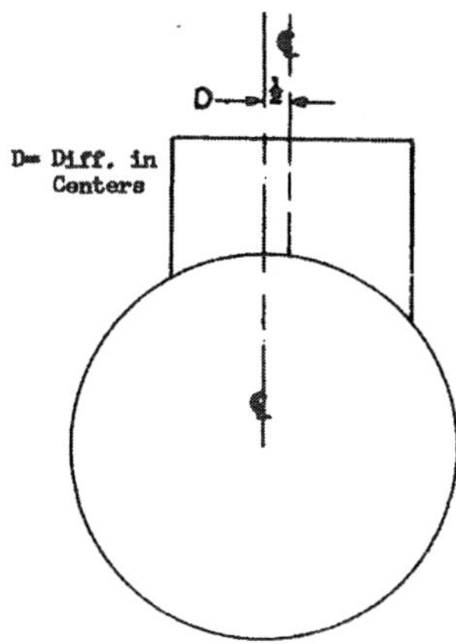

The explanation on the following pages shows how to find eight ordinate lengths. If it is desired to find sixteen ordinate lengths, then the acute angles in the triangle in the I.D. circle will be 67½ degrees and 22½ degrees instead of 45 degrees.

HOW TO FIND EIGHT ORDINATE LENGTHS FOR OFF-CENTER INTERSECTIONS

How to find ordinate number 0 for 6" riser on 8" header:

In this problem, ordinate number 0 has length because the center of the riser and the center of the header are not concentric. By examining the two circles, it can be seen that side DE of triangle DEF is the same length as the difference in centers of the circles. In this example problem, the difference in centers is ½".

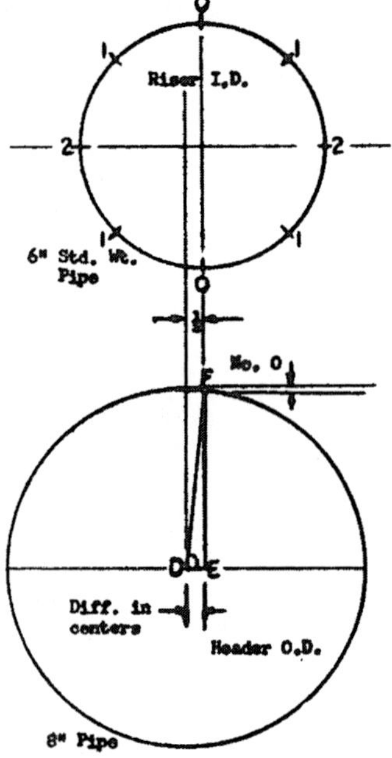

Side DF is the hypotenuse of triangle DEF and its length is one-half the O.D. of the header. Find angle D by dividing the hypotenuse by the adjacent side.

$\frac{\text{Hypotenuse}}{\text{Adjacent}}$ = Secant

$\frac{4.3125}{.5}$ = 8.625 (Look up under secant)

The angle that has 8.6250 for its secant is 83°-21'. This is angle D.

Find side FE of triangle DEF by multiplying the hypotenuse times the sine of 83°-21'.

Opposite = hypotenuse X sine

Opposite = 4.3125 X .99327

Opposite = 4.283" or 4 9/32"
This is side FE.

Subtract FE from one half the OD of the header to obtain length of ordinate number 0.

	4 10/32
−	4 9/32
	1/32" No. 0

For ordinate number 0 there is only 1/32" of drop down on this size pipe.

HOW TO FIND EIGHT ORDINATE LENGTHS FOR OFF-CENTER INTERSECTIONS

How to find ordinate number 1 for 6" riser on 8" header:

By examining the two circles, it can be seen that side DE of triangle DEF is composed of side BC of triangle ABC plus the difference in centers (½"). In triangle ABC, the hypotenuse is one-half of the inside diameter of the riser and angle A is 45 degrees because the circle is divided into eight equal parts.

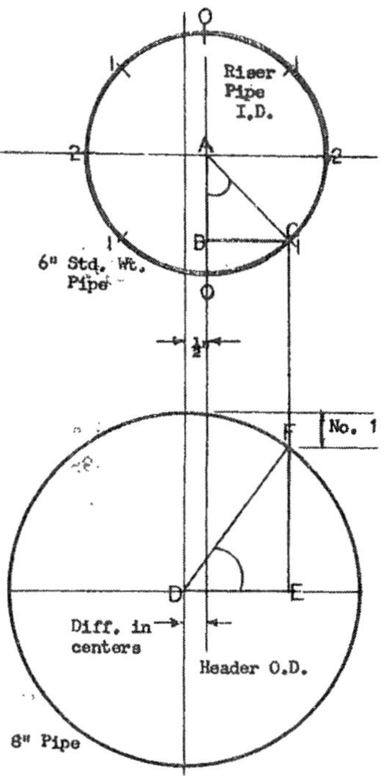

Side BC is the opposite side of angle A. Find BC by multiplying the sine of 45° times the hypotenuse.

Opposite = Hypotenuse X sine
Opposite = 3" X .707
Opposite = 2.121 side BC

Add the difference in centers (½") to BC to find side DE in triangle DEF. It is 2.621 or 2 5/8".

Side DF is the hypotenuse of triangle DEF and its length is one-half of the O.D. of the header. Find angle D by dividing the hypotenuse by the adjacent side.

$\frac{\text{Hypotenuse}}{\text{Adjacent}}$ = Secant

$\frac{4.3125}{2.621}$ = 1.6453 (Look up under secant)

The angle that has 1.6453 for its secant is 52°-34'. This is angle D.

Find side FE of triangle DEF by multiplying the hypotenuse times the sine of 52°-34'.

Opposite = hypotenuse X sine

Opposite = 4.3125" X .79406

Opposite = 3.424" or 3 7/16" which is FE

Subtract FE from one-half the outside diameter of the header to obtain the length of ordinate number 1.

$$\begin{array}{r} 4\ \ 5/16" \\ -\ \underline{3\ \ 7/16"} \\ 7/8" \end{array}$$ Length of number 1 ordinate

HOW TO FIND 8 ORDINATE LENGTHS FOR OFF-CENTER INTERSECTIONS

How to find ordinate # 1 on opposite side of 6" riser.

The only difference in finding this ordinate length and the # 1 previously found is that the difference in centers is <u>subtracted</u> from side CB in triangle ABC to obtain side DE in triangle DEF.

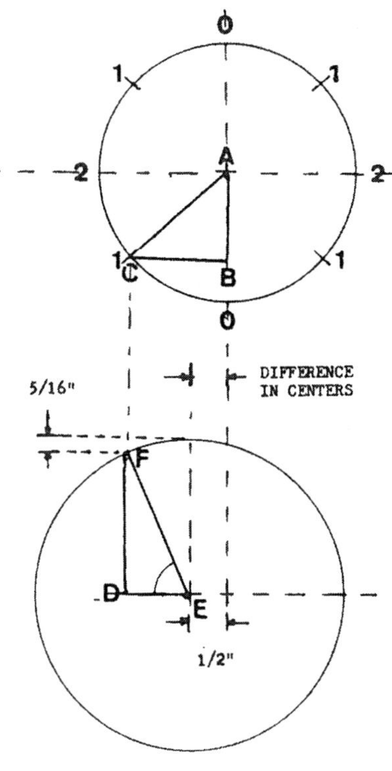

Side CB is the opposite side of angle A in triangle ABC. Find CB by multiplying the sine of 45°times the hypotenuse AC (one-half of I.D.).

Opposite = hypotenuse X sine
Opposite = .707 X 3"
Opposite = 2.121" (side BC)

Subtract difference in centers (½")from CB to find side DE in triangle DEF. It is 1.625".

Side FE is the hypotenuse of triangle DEF and its length is one-half of the O.D. of the header. Find angle E by dividing the hypotenuse by the adjacent side and looking under secant. Angle E is 67°-52'.

Find side FD by multiplying the hypotenuse FE by the sine of 67°-52'.

Opposite = 4.3125 X .92631
Opposite = 3.99" or 4" which is side FD.

Subtract FD from ½ of the O.D. of the header to obtain ordinate # 1. It is 5/16" long. (4 5/16" minus 4" = 5/16")

HOW TO FIND 8 ORDINATE LENGTHS FOR OFF-CENTER INTERSECTIONS

How to find ordinate #2 for 6" riser on 8" header

Side DE of triangle DEF is composed of one-half of the inside diameter of the riser plus the difference in centers (1/2"). Side DE then is 3-1/2" long and is the adjacent side of angle D.

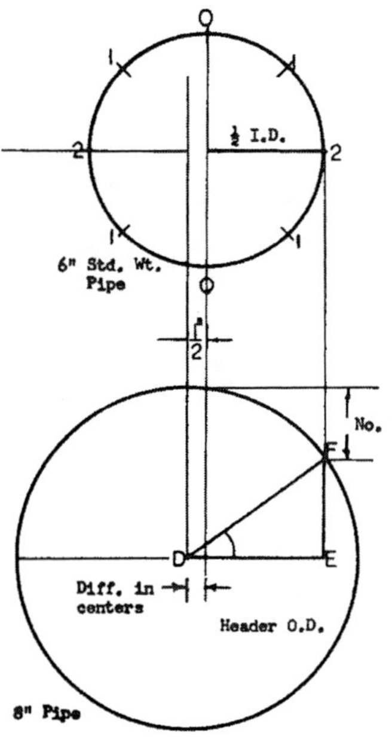

Side DF of triangle DEF is the hypotenuse of the triangle and its length is one-half of the outside diameter of the header. In this example problem it is 4-5/16" long.

Find angle D by dividing the hypotenuse by the adjacent side.

$$\frac{\text{Hypotenuse}}{\text{Adjacent}} = \text{Secant}$$

$$\frac{4.3125}{3.5} = 1.2321$$ (look up under secant)

From the trig tables, the angle that has 1.2321 for its secant is 35°-45'.

Find side FE of triangle DEF by multiplying the hypotenuse times the sine of 35°-45'.

Opposite = hypotenuse X sine
Opposite = 4.3125" X .58425
Opposite = 2.5" or 2-1/2" which is FE

Subtract FE from one-half the outside diameter of the header to obtain the length of ordinate #2. It is 1-13/16" long. (4-5/16" minus 2-1/2" = 1-13/16")

HOW TO FIND 8 ORDINATE LENGTHS FOR OFF-CENTER INTERSECTIONS

How to find # 2 ordinate on opposite side of 6" riser

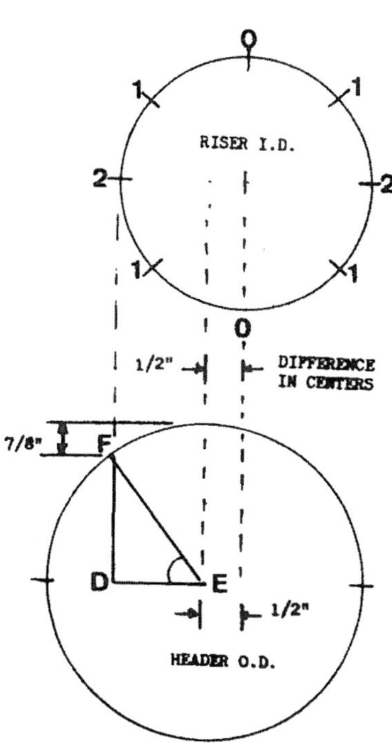

Side DE in triangle DEF is 1/2 of the riser I.D. minus the difference in centers (1/2") It is 2-1/2" long and is the adjacent side of the triangle.

Side FE is 1/2 of the O.D. of the header (4-5/16") and is the hypotenuse of triangle DEF.

Find angle E by dividing the hypotenuse by the adjacent side and looking under secant for the degree.

4.3125 divided by 2.5 is 1.7250 (secant).

The degree that has 1.7250 for its secant is 54°-34'. Find FD by multiplying the sine of 54°-34' by FE.

.81479 X 4.3125 is 3.41" or 3-7/16" which is FD.

Subtract FD from 1/2 of the O.D. of the header to obtain ordinate # 2. It is 7/8" long. (4-5/16")minus 3-7/16" is 7/8".

HOW TO DETERMINE WHERE TO PUT WRAPAROUND LINE ON 45° RISER TO OBTAIN A SPECIFIC DIMENSION FROM CENTER OF HEADER TO END OF RISER

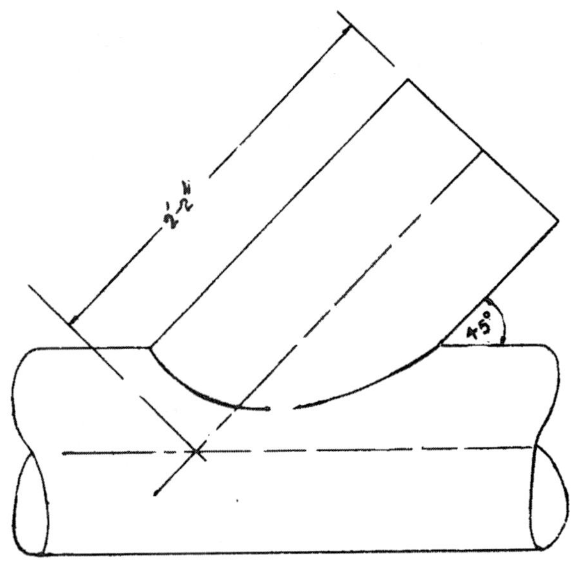

Example Problem: Find the dimension from the end of the riser pipe to the wraparound line. The riser is 8" standard weight pipe saddled on an 8" header.

The wraparound line from which the saddle is laid off must be a specific distance from the end of the riser in order for the 2'-2" center to end measurement to be correct. This dimension must be calculated by solving right triangles as described on the following page.

This is done by finding length AC and subtracting it from the overall length of the riser. This distance is measured from the end of the pipe that is to be the riser and the wraparound line is scribed at that point. When the ordinate lengths for the saddle are laid off from this line, the overall dimension will be 2'-2" when the riser is installed on the header.

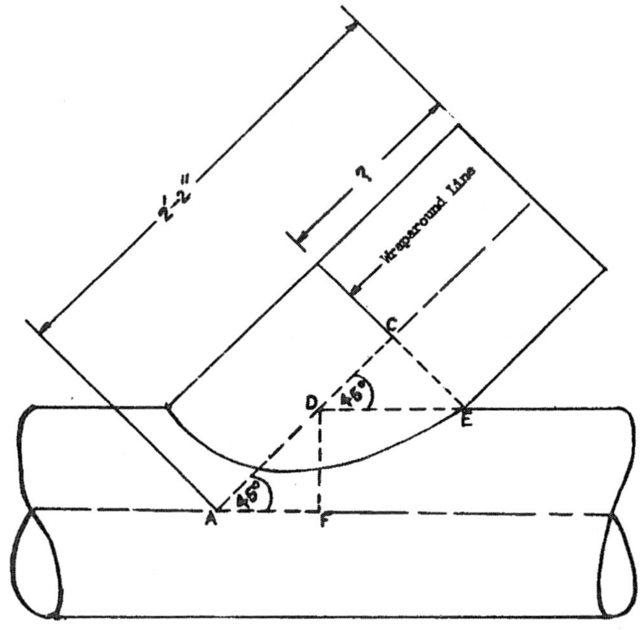

Solution:

Triangles AFD and DCE are both 45° right triangles. Find side AD and DC and add them together to obtain length AC.

In triangle AFD angle A is 45° and the opposite side DF is one-half the O.D. of the header pipe.

Hypotenuse (AD) = Cosecant of angle A times opposite side DF

AD = 1.414 X 4.3125"

AD = 6.098"

Since triangle DCE is a 45° right triangle, the two legs DC and CE are both the same length. CE is one-half the I.D. of the riser pipe, which in this example problem is 4".

```
        AD = 6.098"
   Plus DC = 4.000
            10.098 or 10 1/8"
```

```
        2' - 2"
Minus     10 1/8"
       1' - 3 7/8"
```

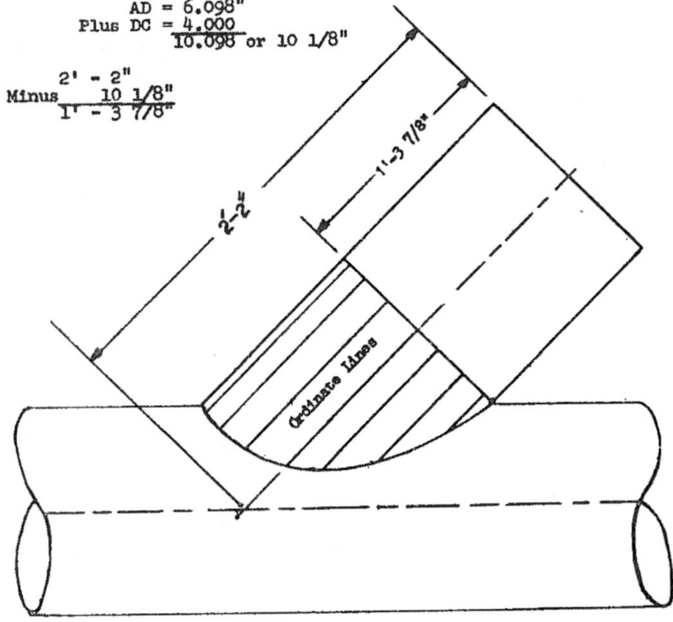

HOW TO DETERMINE WHERE TO PUT WRAPAROUND LINE ON 30° RISER
TO OBTAIN A SPECIFIC DIMENSION FROM CENTER OF HEADER TO END OF RISER

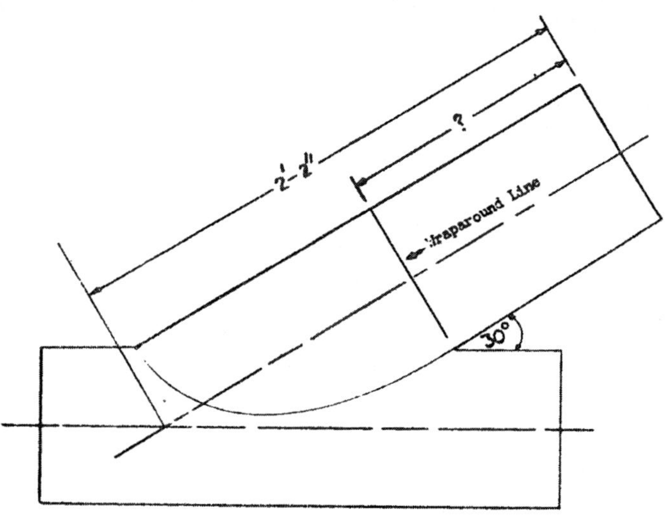

Example Problem:

Find the dimension from the end of the riser pipe to the wraparound line. The riser is 4" standard weight pipe saddled on 4" header.

The wraparound line from which the saddle is laid off must be a specific distance from the end of the riser in order for the 2'-2" center of header to end of riser measurement to be correct. This dimension must be calculated by solving right triangles using the same method previously described for a 45° intersection.

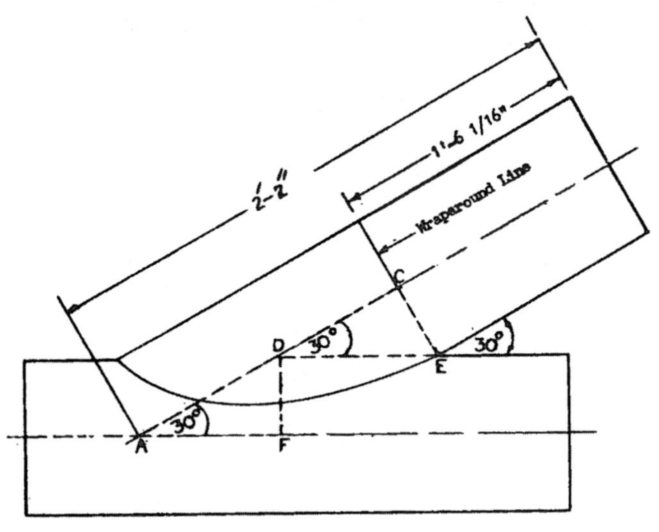

Find the length of hypotenuse AD of triangle AFD and the adjacent side of angle D of triangle DCE. Subtract their sum from the over-all dimension 2'-2" to find out how far from the end of the riser to scribe the wraparound line.

Angle A is 30°. Multiply the cosecant of angle A times ½ of the O.D. of the header pipe (DF) to obtain length of AD.

Hypotenuse = Cosecant times Opposite side
Hypotenuse = 2.000 X 2.25"
Hypotenuse = 4.5" (AD)

Angle D of triangle DCE is also 30°. Multiply the cotangent of angle D times one-half of the I.D. of the riser (CE) to obtain the length of the adjacent side DC.

Adjacent = Cotangent times opposite side
Adjacent = 1.73 X 2"
Adjacent = 3.46" Side DC

```
  4.5"  (AD)
+ 3.46  (DC)
  7.96" Length of side AC
```

```
        2'-2"
Minus       7  15/16"
        1'-6    1/16"   Distance from end of riser to wraparound
                        line.
```

DECIMAL EQUIVALENTS

OF COMMON FRACTIONS OF AN INCH

Fraction	Decimal	Fraction	Decimal
1/64	.015625	33/64	.515625
1/32	.03125	17/32	.53125
3/64	.046875	35/64	.546875
1/16	.0625	9/16	.5625
5/64	.078125	37/64	.578125
3/32	.09375	19/32	.59375
7/64	.109375	39/64	.609375
1/8	.125	5/8	.625
9/64	.140625	41/64	.640625
5/32	.15625	21/32	.65625
11/64	.171875	43/64	.671875
3/16	.1875	11/16	.6875
13/64	.203125	45/64	.703125
7/32	.21875	23/32	.71875
15/64	.234375	47/64	.734375
1/4	.25	3/4	.75
17/64	.265625	49/64	.765625
9/32	.28125	25/32	.78125
19/64	.296875	51/64	.796875
5/16	.3125	13/16	.8125
21/64	.328125	53/64	.828125
11/32	.34375	27/32	.84375
23/64	.359375	55/64	.859375
3/8	.375	7/8	.875
25/64	.390625	57/64	.890625
13/32	.40625	29/32	.90625
27/64	.421875	59/64	.921875
7/16	.4375	15/16	.9375
29/64	.453125	61/64	.953125
15/32	.46875	31/32	.96875
31/64	.484375	63/64	.984375
1/2	.5	1	1.0

TABLES OF TRIGONOMETRIC FUNCTIONS

0°

M	Sine	Cosine	Tan.	Cotan.	Secant	Cosec.	M
0	.00000	1.0000	.00000	Infinite	1.0000	Infinite	60
1	.00029	.0000	.00029	3437.7	.0000	3437.7	59
2	.00058	.0000	.00058	1718.9	.0000	1718.9	58
3	.00087	.0000	.00087	1145.9	.0000	1145.9	57
4	.00116	.0000	.00116	859.44	.0000	859.44	56
5	.00145	1.0000	.00145	687.55	1.0000	687.55	55
6	.00174	.0000	.00174	572.96	.0000	572.96	54
7	.00204	.0000	.00204	491.11	.0000	491.11	53
8	.00233	.0000	.00233	429.72	.0000	429.72	52
9	.00262	.0000	.00262	381.97	.0000	381.97	51
10	.00291	.99999	.00291	343.77	1.0000	343.77	50
11	.00320	.99999	.00320	312.52	.0000	312.52	49
12	.00349	.99999	.00349	286.48	.0000	286.48	48
13	.00378	.99999	.00378	264.44	.0000	264.44	47
14	.00407	.99999	.00407	245.55	.0000	245.55	46
15	.00436	.99999	.00436	229.18	1.0000	229.18	45
16	.00465	.99999	.00465	214.86	.0000	214.86	44
17	.00494	.99999	.00494	202.22	.0000	202.22	43
18	.00524	.99999	.00524	190.98	.0000	190.99	42
19	.00553	.99998	.00553	180.93	.0000	180.93	41
20	.00582	.99998	.00582	171.88	1.0000	171.89	40
21	.00611	.99998	.00611	163.70	.0000	163.70	39
22	.00640	.99998	.00640	156.26	.0000	156.26	38
23	.00669	.99998	.00669	149.46	.0000	149.47	37
24	.00698	.99997	.00698	143.24	.0000	143.24	36
25	.00727	.99997	.00727	137.51	1.0000	137.51	35
26	.00756	.99997	.00756	132.22	.0000	132.22	34
27	.00785	.99997	.00785	127.32	.0000	127.32	33
28	.00814	.99997	.00814	122.77	.0000	122.78	32
29	.00843	.99996	.00844	118.54	.0000	118.54	31
30	.00873	.99996	.00873	114.59	1.0000	114.59	30
31	.00902	.99996	.00902	110.89	.0000	110.90	29
32	.00931	.99996	.00931	107.43	.0000	107.43	28
33	.00960	.99995	.00960	104.17	.0000	104.17	27
34	.00989	.99995	.00989	101.11	.0000	101.11	26
35	.01018	.99995	.01018	98.218	1.0000	98.223	25
36	.01047	.99994	.01047	95.489	.0000	95.495	24
37	.01076	.99994	.01076	92.908	.0000	92.914	23
38	.01105	.99994	.01105	90.463	.0001	90.469	22
39	.01134	.99993	.01134	88.143	.0001	88.149	21
40	.01163	.99993	.01164	85.940	1.0001	85.946	20
41	.01193	.99993	.01193	83.843	.0001	83.849	19
42	.01222	.99992	.01222	81.847	.0001	81.853	18
43	.01251	.99992	.01251	79.943	.0001	79.950	17
44	.01280	.99992	.01280	78.126	.0001	78.133	16
45	.01309	.99991	.01309	76.390	1.0001	76.396	15
46	.01338	.99991	.01338	74.729	.0001	74.736	14
47	.01367	.99991	.01367	73.139	.0001	73.146	13
48	.01396	.99990	.01396	71.615	.0001	71.622	12
49	.01425	.99990	.01425	70.153	.0001	70.160	11
50	.01454	.99989	.01454	68.750	1.0001	68.757	10
51	.01483	.99989	.01484	67.402	.0001	67.409	9
52	.01512	.99988	.01513	66.105	.0001	66.113	8
53	.01542	.99988	.01542	64.858	.0001	64.866	7
54	.01571	.99988	.01571	63.657	.0001	63.664	6
55	.01600	.99987	.01600	62.499	1.0001	62.507	5
56	.01629	.99987	.01629	61.383	.0001	61.391	4
57	.01658	.99987	.01658	60.306	.0001	60.314	3
58	.01687	.99986	.01687	59.266	.0001	59.274	2
59	.01716	.99985	.01716	58.261	.0001	58.270	1
60	.01745	.99985	.01745	57.290	1.0001	57.299	0
M	Cosine	Sine	Cotan.	Tan.	Cosec.	Secant	M

89°

1°

M	Sine	Cosine	Tan.	Cotan.	Secant	Cosec.	M
0	.01745	.99985	.01745	57.290	1.0001	57.299	60
1	.01774	.99984	.01775	56.350	.0001	56.359	59
2	.01803	.99984	.01804	55.441	.0001	55.450	58
3	.01832	.99983	.01833	54.561	.0002	54.570	57
4	.01861	.99983	.01862	53.708	.0002	53.718	56
5	.01891	.99982	.01891	52.882	1.0002	52.891	55
6	.01920	.99981	.01920	52.081	.0002	52.090	54
7	.01949	.99981	.01949	51.303	.0002	51.313	53
8	.01978	.99980	.01978	50.548	.0002	50.558	52
9	.02007	.99980	.02007	49.816	.0002	49.826	51
10	.02036	.99979	.02036	49.104	1.0002	49.114	50
11	.02065	.99979	.02066	48.412	.0002	48.422	49
12	.02094	.99978	.02095	47.739	.0002	47.750	48
13	.02123	.99977	.02124	47.085	.0002	47.096	47
14	.02152	.99977	.02153	46.449	.0002	46.460	46
15	.02181	.99976	.02182	45.829	1.0002	45.840	45
16	.02210	.99975	.02211	45.226	.0002	45.237	44
17	.02240	.99975	.02240	44.638	.0002	44.650	43
18	.02269	.99974	.02269	44.066	.0002	44.077	42
19	.02298	.99974	.02298	43.508	.0003	43.520	41
20	.02326	.99973	.02327	42.964	1.0003	42.976	40
21	.02356	.99972	.02357	42.433	.0003	42.445	39
22	.02385	.99971	.02386	41.916	.0003	41.928	38
23	.02414	.99971	.02415	41.410	.0003	41.423	37
24	.02443	.99970	.02444	40.917	.0003	40.930	36
25	.02472	.99969	.02473	40.436	1.0003	40.448	35
26	.02501	.99969	.02502	39.965	.0003	39.978	34
27	.02530	.99968	.02531	39.506	.0003	39.518	33
28	.02559	.99967	.02560	39.057	.0003	39.069	32
29	.02589	.99966	.02589	38.618	.0003	38.631	31
30	.02618	.99966	.02618	38.188	1.0003	38.201	30
31	.02647	.99965	.02648	37.769	.0003	37.782	29
32	.02676	.99964	.02677	37.358	.0003	37.371	28
33	.02705	.99963	.02706	36.956	.0004	36.969	27
34	.02734	.99963	.02735	36.563	.0004	36.576	26
35	.02763	.99962	.02764	36.177	1.0004	36.191	25
36	.02792	.99961	.02793	35.800	.0004	35.814	24
37	.02821	.99960	.02822	35.431	.0004	35.445	23
38	.02850	.99959	.02851	35.069	.0004	35.084	22
39	.02879	.99958	.02880	34.715	.0004	34.729	21
40	.02908	.99958	.02910	34.368	1.0004	34.382	20
41	.02937	.99957	.02939	34.027	.0004	34.042	19
42	.02967	.99956	.02968	33.693	.0004	33.708	18
43	.02996	.99955	.02997	33.366	.0004	33.381	17
44	.03025	.99954	.03026	33.045	.0004	33.060	16
45	.03054	.99953	.03055	32.730	1.0005	32.745	15
46	.03083	.99952	.03084	32.421	.0005	32.437	14
47	.03112	.99951	.03113	32.118	.0005	32.134	13
48	.03141	.99951	.03143	31.820	.0005	31.836	12
49	.03170	.99950	.03172	31.528	.0005	31.544	11
50	.03199	.99949	.03201	31.241	1.0005	31.257	10
51	.03228	.99948	.03230	30.960	.0005	30.976	9
52	.03257	.99947	.03259	30.683	.0005	30.699	8
53	.03286	.99946	.03288	30.411	.0005	30.428	7
54	.03315	.99945	.03317	30.145	.0005	30.161	6
55	.03344	.99944	.03346	29.882	1.0005	29.899	5
56	.03374	.99943	.03375	29.624	.0006	29.641	4
57	.03403	.99942	.03405	29.371	.0006	29.388	3
58	.03432	.99941	.03434	29.122	.0006	29.139	2
59	.03461	.99940	.03463	28.877	.0006	28.894	1
60	.03490	.99939	.03492	28.636	1.0006	28.654	0
M	Cosine	Sine	Cotan.	Tan.	Cosec.	Secant	M

88°

2°

M	Sine	Cosine	Tan.	Cotan.	Secant	Cosec.	M
0	.03490	.99939	.03492	28.636	1.0006	28.654	60
1	.03519	.99938	.03521	28.399	.0006	28.417	59
2	.03548	.99937	.03550	28.166	.0006	28.184	58
3	.03577	.99936	.03579	27.937	.0006	27.955	57
4	.03606	.99935	.03608	27.712	.0006	27.730	56
5	.03635	.99934	.03638	27.490	1.0007	27.508	55
6	.03664	.99933	.03667	27.271	.0007	27.290	54
7	.03693	.99932	.03696	27.056	.0007	27.075	53
8	.03722	.99931	.03725	26.845	.0007	26.864	52
9	.03751	.99930	.03754	26.637	.0007	26.655	51
10	.03781	.99928	.03783	26.432	1.0007	26.450	50
11	.03810	.99927	.03812	26.230	.0007	26.249	49
12	.03839	.99926	.03842	26.031	.0007	26.050	48
13	.03868	.99925	.03871	25.835	.0007	25.854	47
14	.03897	.99924	.03900	25.642	.0008	25.661	46
15	.03926	.99923	.03929	25.452	1.0008	25.471	45
16	.03955	.99922	.03958	25.264	.0008	25.284	44
17	.03984	.99921	.03987	25.080	.0008	25.100	43
18	.04013	.99919	.04016	24.898	.0008	24.918	42
19	.04042	.99918	.04045	24.718	.0008	24.739	41
20	.04071	.99917	.04075	24.542	1.0008	24.562	40
21	.04100	.99916	.04104	24.367	.0008	24.388	39
22	.04129	.99915	.04133	24.196	.0008	24.216	38
23	.04158	.99913	.04162	24.026	.0009	24.047	37
24	.04187	.99912	.04191	23.859	.0009	23.880	36
25	.04217	.99911	.04220	23.694	1.0009	23.716	35
26	.04246	.99910	.04249	23.532	.0009	23.553	34
27	.04275	.99908	.04279	23.372	.0009	23.393	33
28	.04304	.99907	.04308	23.214	.0009	23.235	32
29	.04333	.99906	.04337	23.058	.0009	23.079	31
30	.04362	.99905	.04366	22.904	1.0009	22.925	30
31	.04391	.99903	.04395	22.752	.0010	22.774	29
32	.04420	.99902	.04424	22.602	.0010	22.624	28
33	.04449	.99901	.04453	22.454	.0010	22.476	27
34	.04478	.99900	.04483	22.308	.0010	22.330	26
35	.04507	.99898	.04512	22.164	1.0010	22.186	25
36	.04536	.99897	.04541	22.022	.0010	22.044	24
37	.04565	.99896	.04570	21.881	.0010	21.904	23
38	.04594	.99894	.04599	21.742	.0010	21.765	22
39	.04623	.99893	.04628	21.606	.0011	21.629	21
40	.04652	.99892	.04657	21.470	1.0011	21.494	20
41	.04681	.99890	.04687	21.337	.0011	21.360	19
42	.04711	.99889	.04716	21.205	.0011	21.228	18
43	.04740	.99888	.04745	21.075	.0011	21.098	17
44	.04769	.99886	.04774	20.946	.0011	20.970	16
45	.04798	.99885	.04803	20.819	1.0011	20.843	15
46	.04827	.99883	.04832	20.693	.0012	20.717	14
47	.04856	.99882	.04862	20.569	.0012	20.593	13
48	.04885	.99881	.04891	20.446	.0012	20.471	12
49	.04914	.99879	.04920	20.325	.0012	20.350	11
50	.04943	.99878	.04949	20.205	1.0012	20.230	10
51	.04972	.99876	.04978	20.087	.0012	20.112	9
52	.05001	.99875	.05007	19.970	.0012	19.995	8
53	.05030	.99873	.05037	19.854	.0013	19.880	7
54	.05059	.99872	.05066	19.740	.0013	19.766	6
55	.05088	.99870	.05095	19.627	1.0013	19.653	5
56	.05117	.99869	.05124	19.515	.0013	19.541	4
57	.05146	.99867	.05153	19.405	.0013	19.431	3
58	.05175	.99866	.05182	19.296	.0013	19.322	2
59	.05204	.99864	.05212	19.188	.0013	19.214	1
60	.05234	.99863	.05241	19.081	1.0014	19.107	0
M	Cosine	Sine	Cotan.	Tan.	Cosec.	Secant	M

87°

3°

M	Sine	Cosine	Tan.	Cotan.	Secant	Cosec.	M
0	.05234	.99863	.05241	19.081	1.0014	19.107	60
1	.05263	.99861	.05270	18.975	.0014	19.002	59
2	.05292	.99860	.05299	18.871	.0014	18.897	58
3	.05321	.99858	.05328	18.768	.0014	18.794	57
4	.05350	.99857	.05357	18.665	.0014	18.692	56
5	.05379	.99855	.05387	18.564	1.0014	18.591	55
6	.05408	.99854	.05416	18.464	.0015	18.491	54
7	.05437	.99852	.05445	18.365	.0015	18.393	53
8	.05466	.99850	.05474	18.268	.0015	18.295	52
9	.05495	.99849	.05503	18.171	.0015	18.198	51
10	.05524	.99847	.05532	18.075	1.0015	18.103	50
11	.05553	.99846	.05562	17.980	.0015	18.008	49
12	.05582	.99844	.05591	17.886	.0016	17.914	48
13	.05611	.99842	.05620	17.793	.0016	17.821	47
14	.05640	.99841	.05649	17.701	.0016	17.730	46
15	.05669	.99839	.05678	17.610	1.0016	17.639	45
16	.05698	.99837	.05707	17.520	.0016	17.549	44
17	.05727	.99836	.05737	17.431	.0016	17.460	43
18	.05756	.99834	.05766	17.343	.0017	17.372	42
19	.05785	.99832	.05795	17.256	.0017	17.285	41
20	.05814	.99831	.05824	17.169	1.0017	17.198	40
21	.05843	.99829	.05853	17.084	.0017	17.113	39
22	.05872	.99827	.05883	16.999	.0017	17.028	38
23	.05902	.99826	.05912	16.915	.0017	16.944	37
24	.05931	.99824	.05941	16.832	.0018	16.861	36
25	.05960	.99822	.05970	16.750	1.0018	16.779	35
26	.05989	.99820	.05999	16.668	.0018	16.698	34
27	.06018	.99819	.06029	16.587	.0018	16.617	33
28	.06047	.99817	.06058	16.507	.0018	16.538	32
29	.06076	.99815	.06087	16.428	.0018	16.459	31
30	.06105	.99813	.06116	16.350	1.0019	16.380	30
31	.06134	.99812	.06145	16.272	.0019	16.303	29
32	.06163	.99810	.06175	16.195	.0019	16.226	28
33	.06192	.99808	.06204	16.119	.0019	16.150	27
34	.06221	.99806	.06233	16.043	.0019	16.075	26
35	.06250	.99804	.06262	15.969	1.0019	16.000	25
36	.06279	.99803	.06291	15.894	.0020	15.926	24
37	.06308	.99801	.06321	15.821	.0020	15.853	23
38	.06337	.99799	.06350	15.748	.0020	15.780	22
39	.06366	.99797	.06379	15.676	.0020	15.708	21
40	.06395	.99795	.06408	15.605	1.0020	15.637	20
41	.06424	.99793	.06437	15.534	.0021	15.566	19
42	.06453	.99791	.06467	15.464	.0021	15.496	18
43	.06482	.99790	.06496	15.394	.0021	15.427	17
44	.06511	.99788	.06525	15.325	.0021	15.358	16
45	.06540	.99786	.06554	15.257	1.0021	15.290	15
46	.06569	.99784	.06583	15.189	.0022	15.222	14
47	.06598	.99782	.06613	15.122	.0022	15.155	13
48	.06627	.99780	.06642	15.056	.0022	15.089	12
49	.06656	.99778	.06671	14.990	.0022	15.023	11
50	.06685	.99776	.06700	14.924	1.0022	14.958	10
51	.06714	.99774	.06730	14.860	.0023	14.893	9
52	.06743	.99772	.06759	14.795	.0023	14.829	8
53	.06772	.99770	.06788	14.732	.0023	14.765	7
54	.06801	.99768	.06817	14.668	.0023	14.702	6
55	.06830	.99766	.06846	14.606	1.0023	14.640	5
56	.06859	.99764	.06876	14.544	.0024	14.578	4
57	.06888	.99762	.06905	14.482	.0024	14.517	3
58	.06918	.99760	.06934	14.421	.0024	14.456	2
59	.06947	.99758	.06963	14.361	.0024	14.395	1
60	.06976	.99756	.06993	14.301	1.0024	14.335	0
M	Cosine	Sine	Cotan.	Tan.	Cosec.	Secant	M

86°

4°

M	Sine	Cosine	Tan.	Cotan.	Secant	Cosec.	M
0	.06976	.99756	.06993	14.301	1.0024	14.335	60
1	.07005	.99754	.07022	14.241	.0025	14.276	59
2	.07034	.99752	.07051	14.182	.0025	14.217	58
3	.07063	.99750	.07080	14.123	.0025	14.159	57
4	.07092	.99748	.07110	14.065	.0025	14.101	56
5	.07121	.99746	.07139	14.008	1.0025	14.043	55
6	.07150	.99744	.07168	13.951	.0026	13.986	54
7	.07179	.99742	.07197	13.894	.0026	13.930	53
8	.07208	.99740	.07226	13.838	.0026	13.874	52
9	.07237	.99738	.07256	13.782	.0026	13.818	51
10	.07266	.99736	.07285	13.727	1.0026	13.763	50
11	.07295	.99733	.07314	13.672	.0027	13.708	49
12	.07324	.99731	.07343	13.617	.0027	13.654	48
13	.07353	.99729	.07373	13.533	.0027	13.600	47
14	.07382	.99727	.07402	13.510	.0027	13.547	46
15	.07411	.99725	.07431	13.457	1.0027	13.494	45
16	.07440	.99723	.07460	13.404	.0028	13.441	44
17	.07469	.99721	.07490	13.351	.0028	13.389	43
18	.07498	.99718	.07519	13.299	.0028	13.337	42
19	.07527	.99716	.07548	13.248	.0028	13.286	41
20	.07556	.99714	.07577	13.197	1.0029	13.235	40
21	.07585	.99712	.07607	13.146	.0029	13.184	39
22	.07614	.99710	.07636	13.096	.0029	13.134	38
23	.07643	.99707	.07665	13.046	.0029	13.084	37
24	.07672	.99705	.07694	12.996	.0029	13.034	36
25	.07701	.99703	.07724	12.947	1.0030	12.985	35
26	.07730	.99701	.07753	12.898	1.0030	12.937	34
27	.07759	.99698	.07782	12.849	.0030	12.888	33
28	.07788	.99696	.07812	12.801	.0030	12.840	32
29	.07817	.99694	.07841	12.754	.0031	12.793	31
30	.07846	.99692	.07870	12.706	1.0031	12.745	30
31	.07875	.99689	.07899	12.659	.0031	12.698	29
32	.07904	.99687	.07929	12.612	.0031	12.652	28
33	.07933	.99685	.07958	12.566	.0032	12.606	27
34	.07962	.99682	.07987	12.520	.0032	12.560	26
35	.07991	.99680	.08016	12.474	1.0032	12.514	25
36	.08020	.99678	.08046	12.429	.0032	12.469	24
37	.08049	.99675	.08075	12.384	.0032	12.424	23
38	.08078	.99673	.08104	12.339	.0033	12.379	22
39	.08107	.99671	.08134	12.295	.0033	12.335	21
40	.08136	.99668	.08163	12.250	1.0033	12.291	20
41	.08165	.99666	.08192	12.207	.0033	12.248	19
42	.08194	.99664	.08221	12.163	.0034	12.204	18
43	.08223	.99661	.08251	12.120	.0034	12.161	17
44	.08252	.99659	.08280	12.077	.0034	12.118	16
45	.08281	.99656	.08309	12.035	1.0034	12.076	15
46	.08310	.99654	.08339	11.992	.0035	12.034	14
47	.08339	.99652	.08368	11.950	.0035	11.992	13
48	.08368	.99649	.08397	11.909	.0035	11.950	12
49	.08397	.99647	.08426	11.867	.0035	11.909	11
50	.08426	.99644	.08456	11.826	1.0036	11.868	10
51	.08455	.99642	.08485	11.785	.0036	11.828	9
52	.08484	.99639	.08514	11.745	.0036	11.787	8
53	.08513	.99637	.08544	11.704	.0036	11.747	7
54	.08542	.99634	.08573	11.664	.0037	11.707	6
55	.08571	.99632	.08602	11.625	1.0037	11.668	5
56	.08600	.99629	.08632	11.585	.0037	11.628	4
57	.08629	.99627	.08661	11.546	1.0037	11.589	3
58	.08658	.99624	.08690	11.507	.0038	11.550	2
59	.08687	.99622	.08719	11.468	.0038	11.512	1
60	.08715	.99619	.08749	11.430	1.0038	11.474	0
M	Cosine	Sine	Cotan.	Tan.	Cosec.	Secant	M

85°

5°

M	Sine	Cosine	Tan.	Cotan.	Secant	Cosec.	M
0	.08715	.99619	.08749	11.430	1.0038	11.474	60
1	.08744	.99617	.08778	11.392	.0038	11.436	59
2	.08773	.99614	.08807	11.354	.0039	11.398	58
3	.08802	.99612	.08837	11.316	.0039	11.360	57
4	.08831	.99609	.08866	11.279	.0039	11.323	56
5	.08860	.99607	.08895	11.242	1.0039	11.286	55
6	.08889	.99604	.08925	11.205	.0040	11.249	54
7	.08918	.99601	.08954	11.168	.0040	11.213	53
8	.08947	.99599	.08983	11.132	.0040	11.176	52
9	.08976	.99596	.09013	11.095	.0040	11.140	51
10	.09005	.99594	.09042	11.059	1.0041	11.104	50
11	.09034	.99591	.09071	11.024	.0041	11.069	49
12	.09063	.99588	.09101	10.988	.0041	11.033	48
13	.09092	.99586	.09130	10.953	.0041	10.998	47
14	.09121	.99583	.09159	10.918	.0042	10.963	46
15	.09150	.99580	.09189	10.883	1.0042	10.929	45
16	.09179	.99578	.09218	10.848	.0042	10.894	44
17	.09208	.99575	.09247	10.814	.0043	10.860	43
18	.09237	.99572	.09277	10.780	.0043	10.826	42
19	.09266	.99570	.09306	10.746	.0043	10.792	41
20	.09295	.99567	.09335	10.712	1.0043	10.758	40
21	.09324	.99564	.09365	10.678	.0044	10.725	39
22	.09353	.99562	.09394	10.645	.0044	10.692	38
23	.09382	.99559	.09423	10.612	.0044	10.659	37
24	.09411	.99556	.09453	10.579	.0044	10.626	36
25	.09440	.99553	.09482	10.546	1.0045	10.593	35
26	.09469	.99551	.09511	10.514	.0045	10.561	34
27	.09498	.99548	.09541	10.481	.0045	10.529	33
28	.09527	.99545	.09570	10.449	.0046	10.497	32
29	.09556	.99542	.09599	10.417	.0046	10.465	31
30	.09584	.99540	.09629	10.385	1.0046	10.433	30
31	.09613	.99537	.09658	10.354	.0046	10.402	29
32	09642	.99534	.09688	10.322	.0047	10.371	28
33	.09671	.99531	.09717	10.291	.0047	10.340	27
34	.09700	.99528	.09746	10.260	.0047	10.309	26
35	.09729	.99525	.09776	10.229	1.0048	10.278	25
36	09758	.99523	09805	10.199	.0048	10.248	24
37	09787	.99520	.09834	10.168	.0048	10.217	23
38	.09816	.99517	.09864	10.138	.0048	10.187	22
39	.09845	.99514	.09893	10.108	.0049	10.157	21
40	.09874	.99511	.09922	10.078	1.0049	10.127	20
41	.09903	.99508	.09952	10 048	.0049	10.098	19
42	.09932	.99505	09981	10.019	.0050	10.068	18
43	.09961	.99503	.10011	9.9893	.0050	10.039	17
44	.09990	.99500	.10040	9.9601	.0050	10.010	16
45	.10019	.99497	10069	9.9310	1.0050	9.9812	15
46	.10048	.99494	.10099	9.9021	.0051	9.9525	14
47	.10077	.99491	.10128	9.8734	.0051	9.9239	13
48	.10106	.99488	.10158	9.8448	.0051	9.8955	12
49	.10134	.99485	.10187	9.8164	0052	9.8672	11
50	.10163	.99482	.10216	9.7882	1.0052	9.8391	10
51	.10192	.99479	.10246	9.7601	.0052	9.8112	9
52	.10221	.99476	.10275	9.7322	.0053	9.7834	8
53	.10250	.99473	.10305	9.7044	.0053	9.7558	7
54	.10279	.99470	.10334	9.6768	0053	9.7283	6
55	.10308	.99467	.10363	9.6493	1.0053	9.7010	5
56	.10337	.99464	10393	9.6220	.0054	9.6739	4
57	.10366	.99461	.10422	9.5949	.0054	9.6469	3
58	.10395	.99458	.10452	9.5679	.0054	9.6200	2
59	.10424	.99455	.10481	9.5411	.0055	9.5933	1
60	.10453	.99452	.10510	9.5144	1.0055	9.5668	0
M	Cosine	Sine	Cotan.	Tan.	Cosec.	Secant	M

84°

6°

M	Sine	Cosine	Tan.	Cotan.	Secant	Cosec.	M
0	.10453	.99452	.10510	9.5144	1.0055	9.5668	60
1	.10482	.99449	.10540	.4878	.0055	.5404	59
2	.10511	.99446	.10569	.4614	0056	.5141	58
3	.10540	.99443	.10599	.4351	.0056	.4880	57
4	.10568	.99440	.10628	.4090	.0056	.4620	56
5	.10597	.99437	.10657	9.3831	1 0057	9.4362	55
6	.10626	.99434	.10687	.3572	.0057	.4105	54
7	.10655	.99431	.10716	.3315	.0057	.3850	53
8	.10684	.99428	.10746	.3060	.0057	.3596	52
9	.10713	.99424	.10775	.2806	.0058	.3343	51
10	.10742	.99421	.10805	9.2553	1.0058	9.3092	50
11	.10771	.99418	.10834	.2302	.0058	.2842	49
12	.10800	.99415	.10863	.2051	.0059	.2593	48
13	.10829	.99412	.10893	.1803	.0059	.2346	47
14	.10858	.99409	.10922	.1555	.0059	.2100	46
15	.10887	.99406	.10952	9.1309	1.0060	9.1855	45
16	.10916	.99402	.10981	.1064	.0060	.1612	44
17	.10944	.99399	.11011	.0821	.0060	.1370	43
18	.10973	.99396	.11040	.0579	.0061	.1129	42
19	.11002	.99393	.11069	.0338	0061	.0890	41
20	.11031	.99390	.11099	9.0098	1.0061	9.0651	40
21	.11060	.99386	.11128	8.9860	.0062	.0414	39
22	.11089	.99383	.11158	.9623	.0062	.0179	38
23	.11118	.99380	.11187	.9387	.0062	8.9944	37
24	.11147	.99377	.11217	.9152	.0063	.9711	36
25	.11176	.99373	.11246	8.8918	1.0063	8 9479	35
26	.11205	.99370	.11276	.8686	.0063	.9248	34
27	.11234	.99367	.11305	.8455	.0064	.9018	33
28	.11262	.99364	.11335	.8225	.0064	.8790	32
29	.11291	.99360	.11364	.7996	.0064	.8563	31
30	.11320	.99357	.11393	8.7769	1.0065	8.8337	30
31	.11349	.99354	.11423	.7542	.0065	.8112	29
32	.11378	.99350	.11452	.7317	.0065	.7888	28
33	.11407	.99347	.11482	.7093	.0066	.7665	27
34	.11436	.99344	.11511	.6870	.0066	.7444	26
35	.11465	.99341	.11541	8.6648	1.0066	8.7223	25
36	.11494	.99337	.11570	.6427	.0067	.7004	24
37	.11523	.99334	.11600	.6208	.0067	.6786	23
38	.11551	.99330	.11629	.5989	.0067	.6569	22
39	.11580	.99327	.11659	.5772	.0068	.6353	21
40	.11609	.99324	.11688	8.5555	1.0068	8.6138	20
41	.11638	.99320	.11718	.5340	.0068	.5924	19
42	.11667	.99317	.11747	.5126	.0069	.5711	18
43	.11696	.99314	.11777	.4913	.0069	.5499	17
44	.11725	.99310	.11806	.4701	.0069	.5289	16
45	.11754	.99307	.11836	8.4489	1.0070	8.5079	15
46	.11783	.99303	.11865	.4279	.0070	.4871	14
47	.11811	.99300	.11895	.4070	.0070	.4663	13
48	.11840	.99296	.11924	.3862	.0071	.4457	12
49	.11869	.99293	.11954	.3655	.0071	.4251	11
50	.11898	.99290	.11983	8.3449	1.0071	8.4046	10
51	.11927	.99286	.12013	.3244	.0072	.3843	9
52	.11956	.99283	.12042	.3040	.0072	.3640	8
53	.11985	.99279	.12072	.2837	.0073	.3439	7
54	.12014	.99276	.12101	.2635	.0073	.3238	6
55	.12042	.99272	.12131	8.2434	1.0073	8.3039	5
56	.12071	.99269	.12160	.2234	.0074	.2840	4
57	.12100	.99265	.12190	.2035	.0074	.2642	3
58	.12129	.99262	.12219	.1837	.0074	.2446	2
59	.12158	.99258	.12249	.1640	.0075	.2250	1
60	.12187	.99255	.12278	8.1443	1.0075	8.2055	0
M	Cosine	Sine	Cotan.	Tan.	Cosec.	Secant	M

83°

7°

M	Sine	Cosine	Tan.	Cotan.	Secant	Cosec.	M
0	.12187	.99255	.12278	8.1443	1.0075	8.2055	60
1	.12216	.99251	.12308	.1248	.0075	.1861	59
2	.12245	.99247	.12337	.1053	.0076	.1668	58
3	.12273	.99244	.12367	.0860	.0076	.1476	57
4	.12302	.99240	.12396	.0667	.0076	.1285	56
5	.12331	.99237	.12426	8.0476	1.0077	8.1094	55
6	.12360	.99233	.12456	.0285	.0077	.0905	54
7	.12389	.99229	.12485	.0095	.0078	.0717	53
8	.12418	.99226	.12515	7.9906	.0078	.0529	52
9	.12447	.99222	.12544	.9717	.0078	.0342	51
10	.12476	.99219	.12574	7.9530	1.0079	8.0156	50
11	.12504	.99215	.12603	.9344	.0079	7.9971	49
12	.12533	.99211	.12633	.9158	.0079	.9787	48
13	.12562	.99208	.12662	.8973	.0080	.9604	47
14	.12591	.99204	.12692	.8789	.0080	.9421	46
15	.12620	.99200	.12722	7.8606	1.0080	7.9240	45
16	.12649	.99197	.12751	.8424	.0081	.9059	44
17	.12678	.99193	.12781	.8243	.0081	.8879	43
18	.12706	.99189	.12810	.8062	.0082	.8700	42
19	.12735	.99186	.12840	.7882	.0082	.8522	41
20	.12764	.99182	.12869	7.7703	1.0082	7.8344	40
21	.12793	.99178	.12899	.7525	.0083	.8168	39
22	.12822	.99174	.12928	.7348	.0083	.7992	38
23	.12851	.99171	.12958	.7171	.0084	.7817	37
24	.12879	.99167	.12988	.6996	.0084	.7642	36
25	.12908	.99163	.13017	7.6821	1.0084	7.7469	35
26	.12937	.99160	.13047	.6646	.0085	.7296	34
27	.12966	.99156	.13076	.6473	.0085	.7124	33
28	.12995	.99152	.13106	.6300	.0085	.6953	32
29	.13024	.99148	.13136	.6129	.0086	.6783	31
30	.13053	.99144	.13165	7.5957	1.0086	7.6613	30
31	.13081	.99141	.13195	.5787	.0087	.6444	29
32	.13110	.99137	.13224	.5617	.0087	.6276	28
33	.13139	.99133	.13254	.5449	.0087	.6108	27
34	.13168	.99129	.13284	.5280	.0088	.5942	26
35	.13197	.99125	.13313	7.5113	1.0088	7.5776	25
36	.13226	.99121	.13343	.4946	.0089	.5611	24
37	.13254	.99118	.13372	.4780	.0089	.5446	23
38	.13283	.99114	.13402	.4615	.0089	.5282	22
39	.13312	.99110	.13432	.4451	.0090	.5119	21
40	.13341	.99106	.13461	7.4287	1.0090	7.4957	20
41	.13370	.99102	.13491	.4124	.0090	.4795	19
42	.13399	.99098	.13520	.3961	.0091	.4634	18
43	.13427	.99094	.13550	.3800	.0091	.4474	17
44	.13456	.99090	.13580	.3639	.0092	.4315	16
45	.13485	.99086	.13609	7.3479	1.0092	7.4156	15
46	.13514	.99083	.13639	.3319	.0092	.3998	14
47	.13543	.99079	.13669	.3160	.0093	.3840	13
48	.13571	.99075	.13698	.3002	.0093	.3683	12
49	.13600	.99071	.13728	.2844	.0094	.3527	11
50	.13629	.99067	.13757	7.2687	1.0094	7.3372	10
51	.13658	.99063	.13787	.2531	.0094	.3217	9
52	.13687	.99059	.13817	.2375	.0095	.3063	8
53	.13716	.99055	.13846	.2220	.0095	.2909	7
54	.13744	.99051	.13876	.2066	.0096	.2757	6
55	.13773	.99047	.13906	7.1912	1.0096	7.2604	5
56	.13802	.99043	.13935	.1759	.0097	.2453	4
57	.13831	.99039	.13965	.1607	.0097	.2302	3
58	.13860	.99035	.13995	.1455	.0097	.2152	2
59	.13888	.99031	.14024	.1304	.0098	.2002	1
60	.13917	.99027	.14054	7.1154	1.0098	7.1853	0
M	Cosine	Sine	Cotan.	Tan.	Cosec.	Secant	M

82°

8°

M	Sine	Cosine	Tan.	Cotan.	Secant	Cosec.	M
0	.13917	.99027	.14054	7.1154	1.0098	7.1853	60
1	.13946	.99023	.14084	.1004	.0099	.1704	59
2	.13975	.99019	.14113	.0854	.0099	.1557	58
3	.14004	.99015	.14143	.0706	.0099	.1409	57
4	.14032	.99010	.14173	.0558	.0100	.1263	56
5	.14061	.99006	.14202	7.0410	1.0100	7.1117	55
6	.14090	.99002	.14232	.0264	.0101	.0972	54
7	.14119	.98998	.14262	.0117	.0101	.0827	53
8	.14148	.98994	.14291	6.9972	.0102	.0683	52
9	.14176	.98990	.14321	.9827	.0102	.0539	51
10	.14205	.98986	.14351	6.9682	1.0102	7.0396	50
11	.14234	.98982	.14380	.9538	.0103	.0254	49
12	.14263	.98978	.14410	.9395	.0103	.0112	48
13	.14292	.98973	.14440	.9252	.0104	6.9971	47
14	.14320	.98969	.14470	.9110	.0104	.9830	46
15	.14349	.98965	.14499	6.8969	1.0104	6.9690	45
16	.14378	.98961	.14529	.8828	.0105	.9550	44
17	.14407	.98957	.14559	.8687	.0105	.9411	43
18	.14436	.98952	.14588	.8547	.0106	.9273	42
19	.14464	.98948	.14618	.8408	.0106	.9135	41
20	.14493	.98944	.14648	6.8269	1.0107	6.8998	40
21	.14522	.98940	.14677	.8131	.0107	.8861	39
22	.14551	.98936	.14707	.7993	.0107	.8725	38
23	.14579	.98931	.14737	.7856	.0108	.8589	37
24	.14608	.98927	.14767	.7720	.0108	.8454	36
25	.14637	.98923	.14796	6.7584	1.0109	6.8320	35
26	.14666	.98919	.14826	.7448	.0109	.8185	34
27	.14695	.98914	.14856	.7313	.0110	.8052	33
28	.14723	.98910	.14886	.7179	.0110	.7919	32
29	.14752	.98906	.14915	.7045	.0111	.7787	31
30	.14781	.98901	.14945	6.6911	1.0111	6.7655	30
31	.14810	.98897	.14975	.6779	.0111	.7523	29
32	.14838	.98893	.15004	.6646	.0112	.7392	28
33	.14867	.98889	.15034	.6514	.0112	.7262	27
34	.14896	.98884	.15064	.6383	.0113	.7132	26
35	.14925	.98880	.15094	6.6252	1.0113	6.7003	25
36	.14953	.98876	.15123	.6122	.0114	.6874	24
37	.14982	.98871	.15153	.5992	.0114	.6745	23
38	.15011	.98867	.15183	.5863	.0115	.6617	22
39	.15040	.98862	.15213	.5734	.0115	.6490	21
40	.15068	.98858	.15243	6.5605	1.0115	6.6363	20
41	.15097	.98854	.15272	.5478	.0116	.6237	19
42	.15126	.98849	.15302	.5350	.0116	.6111	18
43	.15155	.98845	.15332	.5223	.0117	.5985	17
44	.15183	.98840	.15362	.5097	.0117	.5860	16
45	.15212	.98836	.15391	6.4971	1.0118	6.5736	15
46	.15241	.98832	.15421	.4845	.0118	.5612	14
47	.15270	.98827	.15451	.4720	.0119	.5488	13
48	.15298	.98823	.15481	.4596	.0119	.5365	12
49	.15328	.98818	.15511	.4472	.0119	.5243	11
50	.15356	.98814	.15540	6.4348	1.0120	6.5121	10
51	.15385	.98809	.15570	.4225	.0120	.4999	9
52	.15413	.98805	.15600	.4103	.0121	.4878	8
53	.15442	.98800	.15630	.3980	.0121	.4757	7
54	.15471	.98796	.15659	.3859	.0122	.4637	6
55	.15500	.98791	.15689	6.3737	1.0122	6.4517	5
56	.15528	.98787	.15719	.3616	.0123	.4398	4
57	.15557	.98782	.15749	.3496	.0123	.4279	3
58	.15586	.98778	.15779	.3376	.0124	.4160	2
59	.15615	.98773	.15809	.3257	.0124	.4042	1
60	.15643	.98769	.15838	6.3137	1.0125	6.3924	0
M	Cosine	Sine	Cotan.	Tan.	Cosec.	Secant	M

81°

9°

M	Sine	Cosine	Tan.	Cotan.	Secant	Cosec.	M
0	.15643	.98769	.15838	6.3137	1.0125	6.3924	60
1	.15672	.98764	.15868	.3019	.0125	.3807	59
2	.15701	.98760	.15898	.2901	.0125	.3690	58
3	.15730	.98755	.15928	.2783	.0126	.3574	57
4	.15758	.98750	.15958	.2665	.0126	.3458	56
5	.15787	.98746	.15987	6.2548	1.0127	6.3343	55
6	.15816	.98741	.16017	.2432	.0127	.3228	54
7	.15844	.98737	.16047	.2316	.0128	.3113	53
8	.15873	.98732	.16077	.2200	.0128	.2999	52
9	.15902	.98727	.16107	.2085	.0129	.2885	51
10	.15931	.98723	.16137	6.1970	1.0129	6.2772	50
11	.15959	.98718	.16167	.1856	.0130	.2659	49
12	.15988	.98714	.16196	.1742	.0130	.2546	48
13	.16017	.98709	.16226	.1628	.0131	.2434	47
14	.16045	.98704	.16256	.1515	.0131	.2322	46
15	.16074	.98700	.16286	6.1402	1.0132	6.2211	45
16	.16103	.98695	.16316	.1290	.0132	.2100	44
17	.16132	.98690	.16346	.1178	.0133	.1990	43
18	.16160	.98685	.16376	.1066	.0133	.1880	42
19	.16189	.98681	.16405	.0955	.0134	.1770	41
20	.16218	.98676	.16435	6.0844	1.0134	6.1661	40
21	.16246	.98671	.16465	.0734	.0135	.1552	39
22	.16275	.98667	.16495	.0624	.0135	.1443	38
23	.16304	.98662	.16525	.0514	.0136	.1335	37
24	.16333	.98657	.16555	.0405	.0136	.1227	36
25	.16361	.98652	.16585	6.0296	1.0136	6.1120	35
26	.16390	.98648	.16615	.0188	.0137	.1013	34
27	.16419	.98643	.16644	.0080	.0137	.0906	33
28	.16447	.98638	.16674	5.9972	.0138	.0800	32
29	.16476	.98633	.16704	.9865	.0138	.0694	31
30	.16505	.98628	.16734	5.9758	1.0139	6.0588	30
31	.16533	.98624	.16764	.9651	.0139	.0483	29
32	.16562	.98619	.16794	.9545	.0140	.0379	28
33	.16591	.98614	.16824	.9439	.0140	.0274	27
34	.16619	.98609	.16854	.9333	.0141	.0170	26
35	.16648	.98604	.16884	5.9228	1.0141	6.0066	25
36	.16677	.98600	.16914	.9123	.0142	5.9963	24
37	.16705	.98595	.16944	.9019	.0142	.9860	23
38	.16734	.98590	.16973	.8915	.0143	.9758	22
39	.16763	.98585	.17003	.8811	.0143	.9655	21
40	.16791	.98580	.17033	5.8708	1.0144	5.9554	20
41	.16820	.98575	.17063	.8605	.0144	.9452	19
42	.16849	.98570	.17093	.8502	.0145	.9351	18
43	.16878	.98565	.17123	.8400	.0145	.9250	17
44	.16906	.98560	.17153	.8298	.0146	.9150	16
45	.16935	.98556	.17183	5.8196	1.0146	5.9049	15
46	.16964	.98551	.17213	.8095	.0147	.8950	14
47	.16992	.98546	.17243	.7994	.0147	.8850	13
48	.17021	.98541	.17273	.7894	.0148	.8751	12
49	.17050	.98536	.17303	.7794	.0148	.8652	11
50	.17078	.98531	.17333	5.7694	1.0149	5.8554	10
51	.17107	.98526	.17363	.7594	.0150	.8456	9
52	.17136	.98521	.17393	.7495	.0150	.8358	8
53	.17164	.98516	.17423	.7396	.0151	.8261	7
54	.17193	.98511	.17453	.7297	.0151	.8163	6
55	.17221	.98506	.17483	5.7199	1.0152	5.8067	5
56	.17250	.98501	.17513	.7101	.0152	.7970	4
57	.17279	.98496	.17543	.7004	.0153	.7874	3
58	.17307	.98491	.17573	.6906	.0153	.7778	2
59	.17336	.98486	.17603	.6809	.0154	.7683	1
60	.17365	.98481	.17633	5.6713	1.0154	5.7588	0
M	Cosine	Sine	Cotan.	Tan.	Cosec.	Secant	M

80°

10°

M	Sine	Cosine	Tan.	Cotan.	Secant	Cosec.	M
0	.17365	.984[illegible]	.17633	5.6713	1.0154	5.7588	60
1	.17393	.98476	.17663	.6616	.0155	.7493	59
2	.17422	.98471	.17693	.6520	.0155	.7398	58
3	.17451	.98465	.17723	.6425	.0156	.7304	57
4	.17479	.98460	.17753	.6329	.0156	.7210	56
5	.17508	.98455	.17783	5.6234	1.0157	5.7117	55
6	.17537	.98450	.17813	.6140	.0157	.7023	54
7	.17565	.98445	.17843	.6045	.0158	.6930	53
8	.17594	.98440	.17873	.5951	.0158	.6838	52
9	.17622	.98435	.17903	.5857	.0159	.6745	51
10	.17651	.98430	.17933	5.5764	1.0159	5.6653	50
11	.17680	.98425	.17963	.5670	.0160	.6561	49
12	.17708	.98419	.17993	.5578	.0160	.6470	48
13	.17737	.98414	.18023	.5485	.0161	.6379	47
14	.17766	.98409	.18053	.5393	.0162	.6288	46
15	.17794	.98404	.18083	5.5301	1.0162	5.6197	45
16	.17823	.98399	.18113	.5209	.0163	.6107	44
17	.17852	.98394	.18143	.5117	.0163	.6017	43
18	.17880	.98388	.18173	.5026	.0164	.5928	42
19	.17909	.98383	.18203	.4936	.0164	.5838	41
20	.17937	.98378	.18233	5.4845	1.0165	5.5749	40
21	.17966	.98373	.18263	.4755	.0165	.5660	39
22	.17995	.98368	.18293	.4665	.0166	.5572	38
23	.18023	.98362	.18323	.4575	.0166	.5484	37
24	.18052	.98357	.18353	.4486	.0167	.5396	36
25	.18080	.98352	.18383	5.4396	1.0167	5.5308	35
26	.18109	.98347	.18413	.4308	.0168	.5221	34
27	.18138	.98341	.18444	.4219	.0169	.5134	33
28	.18166	.98336	.18474	.4131	.0169	.5047	32
29	.18195	.98331	.18504	.4043	.0170	.4960	31
30	.18223	.98325	.18534	5.3955	1.0170	5.4874	30
31	.18252	.98320	.18564	.3868	.0171	.4788	29
32	.18281	.98315	.18594	.3780	.0171	.4702	28
33	.18309	.98309	.18624	.3694	.0172	.4617	27
34	.18338	.98304	.18654	.3607	.0172	.4532	26
35	.18366	.98299	.18684	5.3521	1.0173	5.4447	25
36	.18395	.98293	.18714	.3434	.0174	.4362	24
37	.18424	.98288	.18745	.3349	.0174	.4278	23
38	.18452	.98283	.18775	.3263	.0175	.4194	22
39	.18481	.98277	.18805	.3178	.0175	.4110	21
40	.18509	.98272	.18835	5.3093	1.0176	5.4026	20
41	.18538	.98267	.18865	.3008	.0176	.3943	19
42	.18567	.98261	.18895	.2923	.0177	.3860	18
43	.18595	.98256	.18925	.2839	.0177	.3777	17
44	.18624	.98250	.18955	.2755	.0178	.3695	16
45	.18652	.98245	.18985	5.2671	1.0179	5.3612	15
46	.18681	.98240	.19016	.2588	.0179	.3530	14
47	.18709	.98234	.19046	.2505	.0180	.3449	13
48	.18738	.98229	.19076	.2422	.0180	.3367	12
49	.18767	.98223	.19106	.2339	.0181	.3286	11
50	.18795	.98218	.19136	5.2257	1.0181	5.3205	10
51	.18824	.98212	.19166	.2174	.0182	.3124	9
52	.18852	.98207	.19197	.2092	.0182	.3044	8
53	.18881	.98201	.19227	.2011	.0183	.2963	7
54	.18909	.98196	.19257	.1929	.0184	.2883	6
55	.18938	.98190	.19287	5.1848	1.0184	5.2803	5
56	.18967	.98185	.19317	.1767	.0185	.2724	4
57	.18995	.96179	.19347	.1686	.0185	.2645	3
58	.19024	.98174	.19378	.1606	.0186	.2566	2
59	.19052	.98168	.19408	.1525	.0186	.2487	1
60	.19081	.98163	.19438	5.1445	1.0187	5.2408	0
M	Cosine	Sine	Cotan.	Tan.	Cosec.	Secant	M

79°

11°

M	Sine	Cosine	Tan.	Cotan.	Secant	Cosec.	M
0	.19081	.98163	.19438	5.1445	1.0187	5.2408	60
1	.19109	.98157	.19468	.1366	.0188	.2330	59
2	.19138	.98152	.19498	.1286	.0188	.2252	58
3	.19166	.98146	.19529	.1207	.0189	.2174	57
4	.19195	.98140	.19559	.1128	.0189	.2097	56
5	.19224	.98135	.19589	5.1049	1.0190	5.2019	55
6	.19252	.98129	.19619	.0970	.0191	.1942	54
7	.19281	.98124	.19649	.0892	.0191	.1865	53
8	.19309	.98118	.19680	.0814	.0192	.1788	52
9	.19338	.98112	.19710	.0736	.0192	.1712	51
10	.19366	.98107	.19740	5.0658	1.0193	5.1636	50
11	.19395	.98101	.19770	.0581	.0193	.1560	49
12	.19423	.98095	.19800	.0504	.0194	.1484	48
13	.19452	.98090	.19831	.0427	.0195	.1409	47
14	.19480	.98084	.19861	.0350	.0195	.1333	46
15	.19509	.98078	.19891	5.0273	1.0196	5.1258	45
16	.19537	.98073	.19921	.0197	.0196	.1183	44
17	.19566	.98067	.19952	.0121	.0197	.1109	43
18	.19595	.98061	.19982	.0045	.0198	.1034	42
19	.19623	.98056	.20012	4.9969	.0198	.0960	41
20	.19652	.98050	.20042	4.9894	1.0199	5.0886	40
21	.19680	.98044	.20073	.9819	.0199	.0812	39
22	.19709	.98039	.20103	.9744	.0200	.0739	38
23	.19737	.98033	.20133	.9669	.0201	.0666	37
24	.19766	.98027	.20163	.9594	.0201	.0593	36
25	.19794	.98021	.20194	4.9520	1.0202	5.0520	35
26	.19823	.98016	.20224	.9446	.0202	.0447	34
27	.19851	.98010	.20254	.9372	.0203	.0375	33
28	.19880	.98004	.20285	.9298	.0204	.0302	32
29	.19908	.97998	.20315	.9225	.0204	.0230	31
30	.19937	.97992	.20345	4.9151	1.0205	5.0158	30
31	.19965	.97987	.20375	.9078	.0205	.0087	29
32	.19994	.97981	.20406	.9006	.0206	.0015	28
33	.20022	.97975	.20436	.8933	.0207	4.9944	27
34	.20051	.97969	.20466	.8860	.0207	.9873	26
35	.20079	.97963	.20497	4.8788	1.0208	4.9802	25
36	.20108	.97957	.20527	.8716	.0208	.9732	24
37	.20136	.97952	.20557	.8644	.0209	.9661	23
38	.20165	.97946	.20588	.8573	.0210	.9591	22
39	.20193	.97940	.20618	.8501	.0210	.9521	21
40	.20222	.97934	.20648	4.8430	1.0211	4.9452	20
41	.20250	.97928	.20679	.8359	.0211	.9382	19
42	.20279	.97922	.20709	.8288	.0212	.9313	18
43	.20307	.97916	.20739	.8217	.0213	.9243	17
44	.20336	.97910	.20770	.8147	.0213	.9175	16
45	.20364	.97904	.20800	4.8077	1.0214	4.9106	15
46	.20393	.97899	.20830	.8007	.0215	.9037	14
47	.20421	.97893	.20861	.7937	.0215	.8969	13
48	.20450	.97887	.20891	.7867	.0216	.8901	12
49	.20478	.97881	.20921	.7798	.0216	.8833	11
50	.20506	.97875	.20952	4.7728	1.0217	4.8765	10
51	.20535	.97869	.20982	.7659	.0218	.8697	9
52	.20563	.97863	.21012	.7591	.0218	.8630	8
53	.20592	.97857	.21043	.7522	.0219	.8563	7
54	.20620	.97851	.21073	.7453	.0220	.8496	6
55	.20649	.97845	.21104	4.7385	1.0220	4.8429	5
56	.20677	.97839	.21134	.7317	.0221	.8362	4
57	.20706	.97833	.21164	.7249	.0221	.8296	3
58	.20734	.97827	.21195	.7181	.0222	.8229	2
59	.20763	.97821	.21225	.7114	.0223	.8163	1
60	.20791	.97815	.21256	4.7046	1.0223	4.8097	0
M	Cosine	Sine	Cotan.	Tan.	Cosec.	Secant	M

78°

12°

M	Sine	Cosine	Tan.	Cotan.	Secant	Cosec.	M
0	.20791	.97815	.21256	4.7046	1.0223	4.8097	60
1	.20820	.97809	.21286	.6979	.0224	8032	59
2	.20848	.97803	.21316	.6912	.0225	.7966	58
3	.20876	.97797	.21347	.6845	.0225	.7901	57
4	.20905	.97790	.21377	.6778	.0226	.7835	56
5	.20933	.97784	.21408	4.6712	1.0226	4.7770	55
6	.20962	.97778	.21438	.6646	.0227	.7706	54
7	.20990	.97772	.21468	.6580	.0228	.7641	53
8	.21019	.97766	21499	.6514	.0228	.7576	52
9	.21047	.97760	.21529	.6448	.0229	.7512	51
10	.21076	.97754	.21560	4.6382	1.0230	4.7448	50
11	.21104	.97748	.21590	.6317	.0230	.7384	49
12	.21132	.97741	.21621	.6252	.0231	.7320	48
13	.21161	.97735	.21651	.6187	.0232	.7257	47
14	.21189	.97729	.21682	.6122	.0232	.7193	46
15	.21218	.97723	.21712	4.6057	1.0233	4.7130	45
16	.21246	.97717	.21742	.5993	.0234	.7067	44
17	.21275	.97711	.21773	.5928	.0234	.7004	43
18	.21303	.97704	.21803	.5864	.0235	.6942	42
19	.21331	.97698	.21834	.5800	.0235	.6879	41
20	.21360	.97692	.21864	4.5736	1.0236	4.6817	40
21	.21388	.97686	.21895	.5673	.0237	.6754	39
22	.21417	.97680	.21925	.5609	.0237	.6692	38
23	.21445	.97673	.21956	.5546	.0238	.6631	37
24	.21473	.97667	.21986	.5483	.0239	.6569	36
25	.21502	.97661	.22017	4.5420	1.0239	4.6507	35
26	.21530	.97655	.22047	.5357	0240	.6446	34
27	.21559	.97648	.22078	.5294	.0241	.6385	33
28	.21587	.97642	.22108	.5232	.0241	.6324	32
29	.21615	.97636	.22139	.5169	.0242	.6263	31
30	.21644	.97630	.22169	4.5107	1.0243	4.6201	30
31	.21672	.97623	.22200	.5045	.0243	.6142	29
32	.21701	.97617	.22230	.4983	.0244	.6081	28
33	.21729	.97611	.22261	.4921	.0245	.6021	27
34	.21757	.97604	.22291	.4860	.0245	5961	26
35	.21786	.97598	.22322	4.4799	1.0246	4.5901	25
36	.21814	.97592	.22353	.4737	.0247	.5841	24
37	.21843	.97585	.22383	.4676	.0247	.5782	23
38	.21871	.97579	.22414	.4615	.0248	.5722	22
39	.21899	.97573	.22444	.4555	.0249	.5663	21
40	.21928	.97566	.22475	4.4494	1.0249	4.5604	20
41	.21956	.97560	.22505	.4434	.0250	.5545	19
42	.21985	.97553	.22536	.4373	.0251	.5486	18
43	.22013	.97547	.22566	.4313	.0251	.5428	17
44	.22041	.97541	.22597	.4253	.0252	.5369	16
45	.22070	.97534	.22628	4.4194	1.0253	4.5311	15
46	.22098	.97528	.22658	.4134	.0253	.5253	14
47	.22126	.97521	.22689	.4074	.0254	5195	13
48	.22155	.97515	.22719	.4015	.0255	.5137	12
49	.22183	.97508	.22750	.3956	.0255	.5079	11
50	.22211	.97502	.22781	4.3897	1.0256	4.5021	10
51	.22240	.97495	.22811	.3838	.0257	.4964	9
52	.22268	.97489	.22842	.3779	.0257	.4907	8
53	.22297	.97483	.22872	.3721	.0258	.4850	7
54	.22325	.97476	.22903	.3662	.0259	.4793	6
55	.22353	.97470	.22934	4.3604	1.0260	4.4736	5
56	.22382	.97463	.22964	.3546	.0260	.4679	4
57	.22410	.97457	.22995	.3488	.0261	.4623	3
58	.22438	.97450	.23025	.3430	.0262	.4566	2
59	.22467	.97443	.23056	.3372	0262	.4510	1
60	.22495	.97437	.23087	4.3315	1.0263	4.4454	0
M	Cosine	Sine	Cotan.	Tan.	Cosec.	Secant	M

77°

13°

M	Sine	Cosine	Tan.	Cotan.	Secant	Cosec.	M
0	.22495	.97437	.23087	4.3315	1.0263	4.4454	60
1	.22523	.97430	.23117	.3257	.0264	.4398	59
2	.22552	.97424	.23148	.3200	.0264	.4342	58
3	.22580	.97417	.23179	.3143	.0265	.4287	57
4	.22608	.97411	.23209	.3086	.0266	.4231	56
5	.22637	.97404	.23240	4.3029	1.0266	4.4176	55
6	.22665	.97398	.23270	.2972	.0267	.4121	54
7	.22693	.97391	.23301	.2916	.0268	.4065	53
8	.22722	.97384	.23332	.2859	.0268	.4011	52
9	.22750	.97378	.23363	.2803	.0269	.3956	51
10	.22778	.97371	.23393	4.2747	1.0270	4.3901	50
11	.22807	.97364	.23424	.2691	.0271	.3847	49
12	.22835	.97358	.23455	.2635	.0271	.3792	48
13	.22863	.97351	.23485	.2579	.0272	.3738	47
14	.22892	.97344	.23516	.2524	.0273	.3684	46
15	.22920	.97338	.23547	4.2468	1.0273	4.3630	45
16	.22948	.97331	.23577	.2413	.0274	.3576	44
17	.22977	.97324	.23608	.2358	.0275	.3522	43
18	.23005	.97318	.23639	.2303	.0276	.3469	42
19	.23033	.97311	.23670	.2248	.0276	.3415	41
20	.23061	.97304	.23700	4.2193	1.0277	4.3362	40
21	.23090	.97298	.23731	.2139	.0278	.3309	39
22	.23118	.97291	.23762	.2084	.0278	.3256	38
23	.23146	.97284	.23793	.2030	.0279	.3203	37
24	.23175	.97277	.23823	.1976	.0280	.3150	36
25	.23203	.97271	.23854	4.1921	1.0280	4.3098	35
26	.23231	.97264	.23885	.1867	.0281	.3045	34
27	.23260	.97257	.23916	.1814	.0282	.2993	33
28	.23288	.97250	.23946	.1760	.0283	.2941	32
29	.23316	.97244	.23977	.1706	.0283	.2888	31
30	.23344	.97237	.24008	4.1653	1.0284	4.2836	30
31	.23373	.97230	.24039	.1600	.0285	.2785	29
32	.23401	.97223	.24069	.1546	.0285	.2733	28
33	.23429	.97216	.24100	.1493	.0286	.2681	27
34	.23458	.97210	.24131	.1440	.0287	.2630	26
35	.23486	.97203	.24162	4.1388	1.0288	4.2579	25
36	.23514	.97196	.24192	.1335	.0288	.2527	24
37	.23542	.97189	.24223	.1282	.0289	.2476	23
38	.23571	.97182	.24254	.1230	.0290	.2425	22
39	.23599	.97175	.24285	.1178	.0291	.2375	21
40	.23627	.97169	.24316	4.1126	1.0291	4.2324	20
41	.23655	.97162	.24346	.1073	.0292	.2273	19
42	.23684	.97155	.24377	.1022	.0293	.2223	18
43	.23712	.97148	.24408	.0970	.0293	.2173	17
44	.23740	.97141	.24439	.0918	.0294	.2122	16
45	.23768	.97134	.24470	4.0867	1.0295	4.2072	15
46	.23797	.97127	.24501	.0815	.0296	.2022	14
47	.23825	.97120	.24531	.0764	.0296	.1972	13
48	.23853	.97113	.24562	.0713	.0297	.1923	12
49	.23881	.97106	.24593	.0662	.0298	.1873	11
50	.23910	.97099	.24624	4.0611	1.0299	4.1824	10
51	.23938	.97092	.24655	.0560	.0299	.1774	9
52	.23966	.97086	.24686	.0509	.0300	.1725	8
53	.23994	.97079	.24717	.0458	.0301	.1676	7
54	.24023	.97072	.24747	.0408	.0302	.1627	6
55	.24051	.97065	.24778	4.0358	1.0302	4.1578	5
56	.24079	.97058	.24809	.0307	.0303	.1529	4
57	.24107	.97051	.24840	.0257	.0304	.1481	3
58	.24136	.97044	.24871	.0207	.0305	.1432	2
59	.24164	.97037	.24902	.0157	.0305	.1384	1
60	.24192	.97029	.24933	4.0108	1.0306	4.1336	0
M	Cosine	Sine	Cotan.	Tan.	Cosec.	Secant	M

76°

14°

M	Sine	Cosine	Tan.	Cotan.	Secant	Cosec.	M
0	.24192	.97029	.24933	4.0108	1.0306	4.1336	60
1	.24220	.97022	.24964	.0058	.0307	.1287	59
2	.24249	.97015	.24995	.0009	.0308	.1239	58
3	.24277	.97008	.25025	3.9959	.0308	.1191	57
4	.24305	.97001	.25056	.9910	.0309	.1144	56
5	.24333	.96994	.25087	3.9861	1.0310	4.1096	55
6	.24361	.96987	.25118	.9812	.0311	.1048	54
7	.24390	.96980	.25149	.9763	.0311	.1001	53
8	.24418	.96973	.25180	.9714	.0312	.0953	52
9	.24446	.96966	.25211	.9665	.0313	.0906	51
10	.24474	.96959	.25242	3.9616	1.0314	4.0859	50
11	.24502	.96952	.25273	.9568	.0314	.0812	49
12	.24531	.96944	.25304	.9520	.0315	.0765	48
13	.24559	.96937	.25335	.9471	.0316	.0718	47
14	.24587	.96930	.25366	.9423	.0317	.0672	46
15	.24615	.96923	.25397	3.9375	1.0317	4.0625	45
16	.24643	.96916	.25428	.9327	.0318	.0579	44
17	.24672	.96909	.25459	.9279	.0319	.0532	43
18	.24700	.96901	.25490	.9231	.0320	.0486	42
19	.24728	.96894	.25521	.9184	.0320	.0440	41
20	.24756	.96887	.25552	3.9136	1.0321	4.0394	40
21	.24784	.96880	.25583	.9089	.0322	.0348	39
22	.24813	.96873	.25614	.9042	.0323	.0302	38
23	.24841	.96865	.25645	.8994	.0323	.0256	37
24	.24869	.96858	.25676	.8947	.0324	.0211	36
25	.24897	.96851	.25707	3.8900	1.0325	4.0165	35
26	.24925	.96844	.25738	.8853	.0326	.0120	34
27	.24953	.96836	.25769	.8807	.0327	.0074	33
28	.24982	.96829	.25800	.8760	.0327	.0029	32
29	.25010	.96822	.25831	.8713	.0328	3.9984	31
30	.25038	.96815	.25862	3.8667	1.0329	3.9939	30
31	.25066	.96807	.25893	.8621	.0330	.9894	29
32	.25094	.96800	.25924	.8574	.0330	.9850	28
33	.25122	.96793	.25955	.8528	.0331	.9805	27
34	.25151	.96785	.25986	.8482	.0332	.9760	26
35	.25179	.96778	.26017	3.8436	1.0333	3.9716	25
36	.25207	.96771	.26048	.8390	.0334	.9672	24
37	.25235	.96763	.26079	.8345	.0334	.9627	23
38	.25263	.96756	.26110	.8299	.0335	.9583	22
39	.25291	.96749	.26141	.8254	.0336	.9539	21
40	.25319	.96741	.26172	3.8208	1.0337	3.9495	20
41	.25348	.96734	.26203	.8163	.0338	.9451	19
42	.25376	.96727	.26234	.8118	.0338	.9408	18
43	.25404	.96719	.26266	.8073	.0339	.9364	17
44	.25432	.96712	.26297	.8027	.0340	.9320	16
45	.25460	.96704	.26328	3.7983	1.0341	3.9277	15
46	.25488	.96697	.26359	.7938	.0341	.9234	14
47	.25516	.96690	.26390	.7893	.0342	.9190	13
48	.25544	.96682	.26421	.7848	.0343	.9147	12
49	.25573	.96675	.26452	.7804	.0344	.9104	11
50	.25601	.96667	.26483	3.7759	1.0345	3.9061	10
51	.25629	.96660	.26514	.7715	.0345	.9018	9
52	.25657	.96652	.26546	.7671	.0346	.8976	8
53	.25685	.96645	.26577	.7627	.0347	.8933	7
54	.25713	.96638	.26608	.7583	.0348	.8890	6
55	.25741	.96630	.26639	3.7539	1.0349	3.8848	5
56	.25769	.96623	.26670	.7495	.0349	.8805	4
57	.25798	.96615	.26701	.7451	.0350	.8763	3
58	.25826	.96608	.26732	.7407	.0351	.8721	2
59	.25854	.96600	.26764	.7364	.0352	.8679	1
60	.25882	.96592	.26795	3.7320	1.0353	3.8637	0
M	Cosine	Sine	Cotan.	Tan.	Cosec.	Secant	M

15°

M	Sine	Cosine	Tan.	Cotan.	Secant	Cosec.	M
0	.25882	.96592	.26795	3.7320	1.0353	3.8637	60
1	.25910	.96585	.26826	.7277	.0353	.8595	59
2	.25938	.96577	.26857	.7234	.0354	.8553	58
3	.25966	.96570	.26888	.7191	.0355	.8512	57
4	.25994	.96562	.26920	.7147	.0356	.8470	56
5	.26022	.96555	.26951	3.7104	1.0357	3.8428	55
6	.26050	.96547	.26982	.7062	.0358	.8387	54
7	.26078	.96540	.27013	.7019	.0358	.8346	53
8	.26107	.96532	.27044	.6976	.0359	.8304	52
9	.26135	.96524	.27076	.6933	.0360	.8263	51
10	.26163	.96517	.27107	3.6891	1.0361	3.8222	50
11	.26191	.96509	.27138	.6848	.0362	.8181	49
12	.26219	.96502	.27169	.6806	.0362	.8140	48
13	.26247	.96494	.27201	.6764	.0363	.8100	47
14	.26275	.96486	.27232	.6722	.0364	.8059	46
15	.26303	.96479	.27263	3.6679	1.0365	3.8018	45
16	.26331	.96471	.27294	.6637	.0366	.7978	44
17	.26359	.96463	.27326	.6596	.0367	.7937	43
18	.26387	.96456	.27357	.6554	.0367	.7897	42
19	.26415	.96448	.27388	.6512	.0368	.7857	41
20	.26443	.96440	.27419	3.6470	1.0369	3.7816	40
21	.26471	.96433	.27451	.6429	.0370	.7776	39
22	.26499	.96425	.27482	.6387	.0371	.7736	38
23	.26527	.96417	.27513	.6346	.0371	.7697	37
24	.26556	.96409	.27544	.6305	.0372	.7657	36
25	.26584	.96402	.27576	3.6263	1.0373	3.7617	35
26	.26612	.96394	.27607	.6222	.0374	.7577	34
27	.26640	.96386	.27638	.6181	.0375	.7538	33
28	.26668	.96378	.27670	.6140	.0376	.7498	32
29	.26696	.96371	.27701	.6100	.0376	.7459	31
30	.26724	.96363	.27732	3.6059	1.0377	3.7420	30
31	.26752	.96355	.27764	.6018	.0378	.7380	29
32	.26780	.96347	.27795	.5977	.0379	.7341	28
33	.26808	.96340	.27826	.5937	.0380	.7302	27
34	.26836	.96332	.27858	.5896	.0381	.7263	26
35	.26864	.96324	.27889	3.5856	1.0382	3.7224	25
36	.26892	.96316	.27920	.5816	.0382	.7186	24
37	.26920	.96308	.27952	.5776	.0383	.7147	23
38	.26948	.96301	.27983	.5736	.0384	.7108	22
39	.26976	.96293	.28014	.5696	.0385	.7070	21
40	.27004	.96285	.28046	3.5656	1.0386	3.7031	20
41	.27032	.96277	.28077	.5616	.0387	.6993	19
42	.27060	.96269	.28109	.5576	.0387	.6955	18
43	.27088	.96261	.28140	.5536	.0388	.6917	17
44	.27116	.96253	.28171	.5497	.0389	.6878	16
45	.27144	.96245	.28203	3.5457	1.0390	3.6840	15
46	.27172	.96238	.28234	.5418	.0391	.6802	14
47	.27200	.96230	.28266	.5378	.0392	.6765	13
48	.27228	.96222	.28297	.5339	.0393	.6727	12
49	.27256	.96214	.28328	.5300	.0393	.6689	11
50	.27284	.96206	.28360	3.5261	1.0394	3.6651	10
51	.27312	.96198	.28391	.5222	.0395	.6614	9
52	.27340	.96190	.28423	.5183	.0396	.6576	8
53	.27368	.96182	.28454	.5144	1.0397	.6539	7
54	.27396	.96174	.28486	.5105	.0398	.6502	6
55	.27424	.96166	.28517	3.5066	1.0399	3.6464	5
56	.27452	.96158	.28549	.5028	.0399	.6427	4
57	.27480	.96150	.28580	.4989	.0400	.6390	3
58	.27508	.96142	.28611	.4951	.0401	.6353	2
59	.27536	.96134	.28643	.4912	.0402	.6316	1
60	.27564	.96126	.28674	3.4874	1.0403	3.6279	0
M	Cosine	Sine	Cotan.	Tan.	Cosec.	Secant	M

74°

16°

M	Sine	Cosine	Tan.	Cotan.	Secant	Cosec.	M
0	.27564	.96126	.28674	3.4874	1.0403	3.6279	60
1	.27592	.96118	.28706	.4836	.0404	.6243	59
2	.27620	.96110	.28737	.4798	.0405	.6206	58
3	.27648	.96102	.28769	.4760	.0406	.6169	57
4	.27675	.96094	.28800	.4722	.0406	.6133	56
5	.27703	.96086	.28832	3.4684	1.0407	3.6096	55
6	.27731	.96078	.28863	.4646	.0408	.6060	54
7	.27759	.96070	.28895	.4608	.0409	.6024	53
8	.27787	.96062	.28926	.4570	.0410	.5987	52
9	.27815	.96054	.28958	.4533	.0411	.5951	51
10	.27843	.96045	.28990	3.4495	1.0412	3.5915	50
11	.27871	.96037	.29021	.4458	.0413	.5879	49
12	.27899	.96029	.29053	.4420	.0413	.5843	48
13	.27927	.96021	.29084	.4383	.0414	.5807	47
14	.27955	.96013	.29116	.4346	.0415	.5772	46
15	.27983	.96005	.29147	3.4308	1.0416	3.5736	45
16	.28011	.95997	.29179	.4271	.0417	.5700	44
17	.28039	.95989	.29210	.4234	.0418	.5665	43
18	.28067	.95980	.29242	.4197	.0419	.5629	42
19	.28094	.95972	.29274	.4160	.0420	.5594	41
20	.28122	.95964	.29305	3.4124	1.0420	3.5559	40
21	.28150	.95956	.29337	.4087	.0421	.5523	39
22	.28178	.95948	.29368	.4050	.0422	.5488	38
23	.28206	.95940	.29400	.4014	.0423	.5453	37
24	.28234	.95931	.29432	.3977	.0424	.5418	36
25	.28262	.95923	.29463	3.3941	1.0425	3.5383	35
26	.28290	.95915	.29495	.3904	.0426	.5348	34
27	.28318	.95907	.29526	.3868	.0427	.5313	33
28	.28346	.95898	.29558	.3832	.0428	.5279	32
29	.28374	.95890	.29590	.3795	.0428	.5244	31
30	.28401	.95882	.29621	3.3759	1.0429	3.5209	30
31	.28429	.95874	.29653	.3723	.0430	.5175	29
32	.28457	.95865	.29685	.3687	.0431	.5140	28
33	.28485	.95857	.29716	.3651	.0432	.5106	27
34	.28513	.95849	.29748	.3616	.0433	.5072	26
35	.28541	.95840	.29780	3.3580	1.0434	3.5037	25
36	.28569	.95832	.29811	.3544	.0435	.5003	24
37	.28597	.95824	.29843	.3509	.0436	.4969	23
38	.28624	.95816	.29875	.3473	.0437	3.4935	22
39	.28652	.95807	.29906	.3438	.0438	.4901	21
40	.28680	.95799	.29938	3.3402	1.0438	3.4867	20
41	.28708	.95791	.29970	.3367	.0439	.4833	19
42	.28736	.95782	.30001	.3332	.0440	.4799	18
43	.28764	.95774	.30033	.3296	.0441	.4766	17
44	.28792	.95765	.30065	.3261	.0442	.4732	16
45	.28820	.95757	.30096	3.3226	1.0443	3.4698	15
46	.28847	.95749	.30128	.3191	.0444	.4665	14
47	.28875	.95740	.30160	.3156	.0445	.4632	13
48	.28903	.95732	.30192	.3121	.0446	.4598	12
49	.28931	.95723	.30223	.3087	.0447	.4565	11
50	.28959	.95715	.30255	3.3052	1.0448	3.4532	10
51	.28987	.95707	.30287	.3017	.0448	.4498	9
52	.29014	.95698	.30319	3.2983	.0449	.4465	8
53	.29042	.95690	.30350	.2948	.0450	.4432	7
54	.29070	.95681	.30382	.2914	.0451	.4399	6
55	.29098	.95673	.30414	3.2879	1.0452	3.4366	5
56	.29126	.95664	.30446	.2845	.0453	.4334	4
57	.29154	.95656	.30478	.2811	.0454	.4301	3
58	.29181	.95647	.30509	.2777	.0455	.4268	2
59	.29209	.95639	.30541	.2742	.0456	.4236	1
60	.29237	.95630	.30573	3.2708	1.0457	3.4203	0
M	Cosine	Sine	Cotan.	Tan.	Cosec.	Secant	M

73°

17°

M	Sine	Cosine	Tan.	Cotan.	Secant	Cosec.	M
0	.29237	.95630	.30573	3.2708	1.0457	3.4203	60
1	.29265	.95622	.30605	.2674	.0458	.4170	59
2	.29293	.95613	.30637	.2640	.0459	.4138	58
3	.29321	.95605	.30668	.2607	.0460	.4106	57
4	.29348	.95596	.30700	.2573	.0461	.4073	56
5	.29376	.95588	.30732	3.2539	1.0461	3.4041	55
6	.29404	.95579	.30764	.2505	.0462	.4009	54
7	.29432	.95571	.30796	.2472	.0463	.3977	53
8	.29460	.95562	.30828	.2438	.0464	.3945	52
9	.29487	.95554	.30859	.2405	.0465	.3913	51
10	.29515	.95545	.30891	3.2371	1.0466	3.3881	50
11	.29543	.95536	.30923	.2338	.0467	.3849	49
12	.29571	.95528	.30955	.2305	.0468	.3817	48
13	.29598	.95519	.30987	.2271	.0469	.3785	47
14	.29626	.95511	.31019	.2238	.0470	.3754	46
15	.29654	.95502	.31051	3.2205	1.0471	3.3722	45
16	.29682	.95493	.31083	.2172	.0472	.3690	44
17	.29710	.95485	.31115	.2139	.0473	.3659	43
18	.29737	.95476	.31146	.2106	.0474	.3627	42
19	.29765	.95467	.31178	.2073	.0475	.3596	41
20	.29793	.95459	.31210	3.2041	1.0476	3.3565	40
21	.29821	.95450	.31242	.2008	.0477	.3534	39
22	.29848	.95441	.31274	.1975	.0478	.3502	38
23	.29876	.95433	.31306	.1942	.0478	.3471	37
24	.29904	.95424	.31338	.1910	.0479	.3440	36
25	.29932	.95415	.31370	3.1877	1.0480	3.3409	35
26	.29959	.95407	.31402	.1845	.0481	.3378	34
27	.29987	.95398	.31434	.1813	.0482	.3347	33
28	.30015	.95389	.31466	.1780	.0483	.3316	32
29	.30043	.95380	.31498	.1748	.0484	.3286	31
30	.30070	.95372	.31530	3.1716	1.0485	3.3255	30
31	.30098	.95363	.31562	.1684	.0486	.3224	29
32	.30126	.95354	.31594	.1652	.0487	.3194	28
33	.30154	.95345	.31626	.1620	.0488	.3163	27
34	.30181	.95337	.31658	.1588	.0489	.3133	26
35	.30209	.95328	.31690	3.1556	1.0490	3.3102	25
36	.30237	.95319	.31722	.1524	.0491	.3072	24
37	.30265	.95310	.31754	.1492	.0492	.3042	23
38	.30292	.95301	.31786	.1460	.0493	.3011	22
39	.30320	.95293	.31818	.1429	.0494	.2981	21
40	.30348	.95284	.31850	3.1397	1.0495	3.2951	20
41	.30375	.95275	.31882	.1366	.0496	.2921	19
42	.30403	.95266	.31914	.1334	.0497	.2891	18
43	.30431	.95257	.31946	.1303	.0498	.2861	17
44	.30459	.95248	.31978	.1271	.0499	.2831	16
45	.30486	.95239	.32010	3.1240	1.0500	3.2801	15
46	.30514	.95231	.32042	.1209	.0501	.2772	14
47	.30542	.95222	.32074	.1177	.0502	.2742	13
48	.30569	.95213	.32106	.1146	.0503	.2712	12
49	.30597	.95204	.32138	.1115	.0504	.2683	11
50	.30625	.95195	.32171	3.1084	1.0505	3.2653	10
51	.30653	.95186	.32203	.1053	.0506	.2624	9
52	.30680	.95177	.32235	.1022	.0507	.2594	8
53	.30708	.95168	.32267	.0991	.0508	.2565	7
54	.30736	.95159	.32299	.0960	.0509	.2535	6
55	.30763	.95150	.32331	3.0930	1.0510	3.2506	5
56	.30791	.95141	.32363	.0899	.0511	.2477	4
57	.30819	.95132	.32395	.0868	.0512	.2448	3
58	.30846	.95124	.32428	.0838	.0513	.2419	2
59	.30874	.95115	.32460	.0807	.0514	.2390	1
60	.30902	.95106	.32492	3.0777	1.0515	3.2361	0
M	Cosine	Sine	Cotan.	Tan.	Cosec.	Secant	M

72°

18°

M	Sine	Cosine	Tan.	Cotan.	Secant	Cosec.	M
0	.30902	.95106	.32492	3.0777	1.0515	3.2361	60
1	.30929	.95097	.32524	.0746	.0516	.2332	59
2	.30957	.95088	.32556	.0716	.0517	.2303	58
3	.30985	.95079	.32588	.0686	.0518	.2274	57
4	.31012	.95070	.32621	.0655	.0519	.2245	56
5	.31040	.95061	.32653	3.0625	1.0520	3.2216	55
6	.31068	.95051	.32685	.0595	.0521	.2188	54
7	.31095	.95042	.32717	.0565	.0522	.2159	53
8	.31123	.95033	.32749	.0535	.0523	.2131	52
9	.31150	.95024	.32782	.0505	.0524	.2102	51
10	.31178	.95015	.32814	3.0475	1.0525	3.2074	50
11	.31206	.95006	.32846	.0445	.0526	.2045	49
12	.31233	.94997	.32878	.0415	.0527	.2017	48
13	.31261	.94988	.32910	.0385	.0528	.1989	47
14	.31289	.94979	.32943	.0356	.0529	.1960	46
15	.31316	.94970	.32975	3.0326	1.0530	3.1932	45
16	.31344	.94961	.33007	.0296	.0531	.1904	44
17	.31372	.94952	.33039	.0267	.0532	.1876	43
18	.31399	.94942	.33072	.0237	.0533	.1848	42
19	.31427	.94933	.33104	.0208	.0534	.1820	41
20	.31454	.94924	.33136	3.0178	1.0535	3.1792	40
21	.31482	.94915	.33169	.0149	.0536	.1764	39
22	.31510	.94906	.33201	.0120	.0537	.1736	38
23	.31537	.94897	.33233	.0090	.0538	.1708	37
24	.31565	.94888	.33265	.0061	.0539	.1681	36
25	.31592	.94878	.33298	3.0032	1.0540	3.1653	35
26	.31620	.94869	.33330	.0003	.0541	.1625	34
27	.31648	.94860	.33362	2.9974	.0542	.1598	33
28	.31675	.94851	.33395	.9945	.0543	.1570	32
29	.31703	.94841	.33427	.9916	.0544	.1543	31
30	.31730	.94832	.33459	2.9887	1.0545	3.1515	30
31	.31758	.94823	.33492	.9858	.0546	.1488	29
32	.31786	.94814	.33524	.9829	.0547	.1461	28
33	.31813	.94805	.33557	.9800	.0548	.1433	27
34	.31841	.94795	.33589	.9772	.0549	.1406	26
35	.31868	.94786	.33621	2.9743	1.0550	3.1379	25
36	.31896	.94777	.33654	.9714	.0551	.1352	24
37	.31923	.94767	.33686	.9686	.0552	.1325	23
38	.31951	.94758	.33718	.9657	.0553	.1298	22
39	.31978	.94749	.33751	.9629	.0554	.1271	21
40	.32006	.94740	.33783	2.9600	1.0555	3.1244	20
41	.32034	.94730	.33816	.9572	.0556	.1217	19
42	.32061	.94721	.33848	.9544	.0557	.1190	18
43	.32089	.94712	.33880	.9515	.0558	.1163	17
44	.32116	.94702	.33913	.9487	.0559	.1137	16
45	.32144	.94693	.33945	2.9459	1.0560	3.1110	15
46	.32171	.94684	.33978	.9431	.0561	.1083	14
47	.32199	.94674	.34010	.9403	.0562	.1057	13
48	.32226	.94665	.34043	.9375	.0563	.1030	12
49	.32254	.94655	.34075	.9347	.0565	.1004	11
50	.32282	.94646	.34108	2.9319	1.0566	3.0977	10
51	.32309	.94637	.34140	.9291	.0567	.0951	9
52	.32337	.94627	.34173	.9263	.0568	.0925	8
53	.32364	.94618	.34205	.9235	.0569	.0898	7
54	.32392	.94608	.34238	.9208	.0570	.0872	6
55	.32419	.94599	.34270	2.9180	1.0571	3.0846	5
56	.32447	.94590	.34303	.9152	.0572	.0820	4
57	.32474	.94580	.34335	.9125	.0573	.0793	3
58	.32502	.94571	.34368	.9097	.0574	.0767	2
59	.32529	.94561	.34400	.9069	.0575	.0741	1
60	.32557	.94552	.34433	2.9042	1.0576	3.0715	0
M	Cosine	Sine	Cotan.	Tan.	Cosec.	Secant	M

71°

19°

M	Sine	Cosine	Tan.	Cotan.	Secant	Cosec.	M
0	.32557	.94552	.34433	2.9042	1.0576	3.0715	60
1	.32584	.94542	.34465	.9015	.0577	.0690	59
2	.32612	.94533	.34498	.8987	.0578	.0664	58
3	.32639	.94523	.34530	.8960	.0579	.0638	57
4	.32667	.94514	.34563	.8933	.0580	.0612	56
5	.32694	.94504	.34595	2.8905	1.0581	3.0586	55
6	.32722	.94495	.34628	.8878	.0582	.0561	54
7	.32749	.94485	.34661	.8851	.0584	.0535	53
8	.32777	.94476	.34693	.8824	.0585	.0509	52
9	.32804	.94466	.34726	.8797	.0586	.0484	51
10	.32832	.94457	.34758	2.8770	1.0587	3.0458	50
11	.32859	.94447	.34791	.8743	.0588	.0433	49
12	.32887	.94438	.34824	.8716	.0589	.0407	48
13	.32914	.94428	.34856	.8689	.0590	.0382	47
14	.32942	.94418	.34889	.8662	.0591	.0357	46
15	.32969	.94409	.34921	2.8636	1.0592	3.0331	45
16	.32996	.94399	.34954	.8609	.0593	.0306	44
17	.33024	.94390	.34987	.8582	.0594	.0281	43
18	.33051	.94380	.35019	.8555	.0595	.0256	42
19	.33079	.94370	.35052	.8529	.0596	.0231	41
20	.33106	.94361	.35085	2.8502	1.0598	3.0206	40
21	.33134	.94351	.35117	.8476	.0599	.0181	39
22	.33161	.94341	.35150	.8449	.0600	.0156	38
23	.33189	.94332	.35183	.8423	.0601	.0131	37
24	.33216	.94322	.35215	.8396	.0602	.0106	36
25	.33243	.94313	.35248	2.8370	1.0603	3.0081	35
26	.33271	.94303	.35281	.8344	.0604	.0056	34
27	.33298	.94293	.35314	.8318	.0605	.0031	33
28	.33326	.94283	.35346	.8291	.0606	.0007	32
29	.33353	.94274	.35379	.8265	.0607	2.9982	31
30	.33381	.94264	.35412	2.8239	1.0608	2.9957	30
31	.33408	.94254	.35445	.8213	.0609	.9933	29
32	.33435	.94245	.35477	.8187	.0611	.9908	28
33	.33463	.94235	.35510	.8161	.0612	.9884	27
34	.33490	.94225	.35543	.8135	.0613	.9859	26
35	.33518	.94215	.35576	2.8109	1.0614	2.9835	25
36	.33545	.94206	.35608	.8083	.0615	.9810	24
37	.33572	.94196	.35641	.8057	.0616	.9786	23
38	.33600	.94186	.35674	.8032	.0617	.9762	22
39	.33627	.94176	.35707	.8006	.0618	.9738	21
40	.33655	.94167	.35739	2.7980	1.0619	2.9713	20
41	.33682	.94157	.35772	.7954	.0620	.9689	19
42	.33709	.94147	.35805	.7929	.0622	.9665	18
43	.33737	.94137	.35838	.7903	.0623	.9641	17
44	.33764	.94127	.35871	.7878	.0624	.9617	16
45	.33792	.94118	.35904	2.7852	1.0625	2.9593	15
46	.33819	.94108	.35936	.7827	.0626	.9569	14
47	.33846	.94098	.35969	.7801	.0627	.9545	13
48	.33874	.94088	.36002	.7776	.0628	.9521	12
49	.33901	.94078	.36035	.7751	.0629	.9497	11
50	.33928	.94068	.36068	2.7725	1.0630	2.9474	10
51	.33956	.94058	.36101	.7700	.0632	.9450	9
52	.33983	.94049	.36134	.7675	.0633	.9426	8
53	.34011	.94039	.36167	.7650	.0634	.9402	7
54	.34038	.94029	.36199	.7625	.0635	.9379	6
55	.34065	.94019	.36232	2.7600	1.0636	2.9355	5
56	.34093	.94009	.36265	.7575	.0637	.9332	4
57	.34120	.93999	.36298	.7550	.0638	.9308	3
58	.34147	.93989	.36331	.7525	.0639	.9285	2
59	.34175	.93979	.36364	.7500	.0641	.9261	1
60	.34202	.93969	.36397	2.7475	1.0642	2.9238	0
M	Cosine	Sine	Cotan.	Tan.	Cosec.	Secant	M

70°

20°

M	Sine	Cosine	Tan.	Cotan.	Secant	Cosec.	M
0	.34202	.93969	.36397	2.7475	1.0642	2.9238	60
1	.34229	.93959	.36430	.7450	.0643	.9215	59
2	.34257	.93949	.36463	.7425	.0644	.9191	58
3	.34284	.93939	.36496	.7400	.0645	.9168	57
4	.34311	.93929	.36529	.7376	.0646	.9145	56
5	.34339	.93919	.36562	2.7351	1.0647	2.9122	55
6	.34366	.93909	.36595	.7326	.0648	.9098	54
7	.34393	.93899	.36628	.7302	.0650	.9075	53
8	.34421	.93889	.36661	.7277	.0651	.9052	52
9	.34448	.93879	.36694	.7252	.0652	.9029	51
10	.34475	.93869	.36727	2.7228	1.0653	2.9006	50
11	.34502	.93859	.36760	.7204	.0654	.8983	49
12	.34530	.93849	.36793	.7179	.0655	.8960	48
13	.34557	.93839	.36826	.7155	.0656	.8937	47
14	.34584	.93829	.36859	.7130	.0658	.8915	46
15	.34612	.93819	.36892	2.7106	1.0659	2.8892	45
16	.34639	.93809	.36925	.7082	.0660	.8869	44
17	.34666	.93799	.36958	.7058	.0661	.8846	43
18	.34693	.93789	.36991	.7033	.0662	.8824	42
19	.34721	.93779	.37024	.7009	.0663	.8801	41
20	.34748	.93769	.37057	2.6985	1.0664	2.8778	40
21	.34775	.93758	.37090	.6961	.0666	.8756	39
22	.34803	.93748	.37123	.6937	.0667	.8733	38
23	.34830	.93738	.37156	.6913	.0668	.8711	37
24	.34857	.93728	.37190	.6889	.0669	.8688	36
25	.34884	.93718	.37223	2.6865	1.0670	2.8666	35
26	.34912	.93708	.37256	.6841	.0671	.8644	34
27	.34939	.93698	.37289	.6817	.0673	.8621	33
28	.34966	.93687	.37322	.6794	.0674	.8599	32
29	.34993	.93677	.37355	.6770	.0675	.8577	31
30	.35021	.93667	.37388	2.6746	1.0676	2.8554	30
31	.35048	.93657	.37422	.6722	.0677	.8532	29
32	.35075	.93647	.37455	.6699	.0678	.8510	28
33	.35102	.93637	.37488	.6675	.0679	.8488	27
34	.35130	.93626	.37521	.6652	.0681	.8466	26
35	.35157	.93616	.37554	2.6628	1.0682	2.8444	25
36	.35184	.93606	.37587	.6604	.0683	.8422	24
37	.35211	.93596	.37621	.6581	.0684	.8400	23
38	.35239	.93585	.37654	.6558	.0685	.8378	22
39	.35266	.93575	.37687	.6534	.0686	.8356	21
40	.35293	.93565	.37720	2.6511	1.0688	2.8334	20
41	.35320	.93555	.37754	.6487	.0689	.8312	19
42	.35347	.93544	.37787	.6464	.0690	.8290	18
43	.35375	.93534	.37820	.6441	.0691	.8269	17
44	.35402	.93524	.37853	.6418	.0692	.8247	16
45	.35429	.93513	.37887	2.6394	1.0694	2.8225	15
46	.35456	.93503	.37920	.6371	.0695	.8204	14
47	.35483	.93493	.37953	.6348	.0696	.8182	13
48	.35511	.93482	.37986	.6325	.0697	.8160	12
49	.35538	.93472	.38020	.6302	.0698	.8139	11
50	.35565	.93462	.38053	2.6279	1.0699	2.8117	10
51	.35592	.93451	.38086	.6256	.0701	.8096	9
52	.35619	.93441	.38120	.6233	.0702	.8074	8
53	.35647	.93431	.38153	.6210	.0703	.8053	7
54	.35674	.93420	.38186	.6187	.0704	.8032	6
55	.35701	.93410	.38220	2.6164	1.0705	2.8010	5
56	.35728	.93400	.38253	.6142	.0707	.7989	4
57	.35755	.93389	.38286	.6119	.0708	.7968	3
58	.35782	.93379	.38320	.6096	.0709	.7947	2
59	.35810	.93368	.38353	.6073	.0710	.7925	1
60	.35837	.93358	.38386	2.6051	1.0711	2.7904	0
M	Cosine	Sine	Cotan.	Tan.	Cosec.	Secant	M

69°

21°

M	Sine	Cosine	Tan.	Cotan.	Secant	Cosec.	M
0	.35837	.93358	.38386	2.6051	1.0711	2.7904	60
1	.35864	.93348	.38420	.6028	.0713	.7883	59
2	.35891	.93337	.38453	.6006	.0714	.7862	58
3	.35918	.93327	.38486	.5983	.0715	.7841	57
4	.35945	.93316	.38520	.5960	.0716	.7820	56
5	.35972	.93306	.38553	2.5938	1.0717	2.7799	55
6	.36000	.93295	.38587	.5916	.0719	.7778	54
7	.36027	.93285	.38620	.5893	.0720	.7757	53
8	.36054	.93274	.38654	.5871	.0721	.7736	52
9	.36081	.93264	.38687	.5848	.0722	.7715	51
10	.36108	.93253	.38720	2.5826	1.0723	2.7694	50
11	.36135	.93243	.38754	.5804	.0725	.7674	49
12	.36162	.93232	.38787	.5781	.0726	.7653	48
13	.36189	.93222	.38821	.5759	.0727	.7632	47
14	.36217	.93211	.38854	.5737	.0728	.7611	46
15	.36244	.93201	.38888	2.5715	1.0729	2.7591	45
16	.36271	.93190	.38921	.5693	.0731	.7570	44
17	.36298	.93180	.38955	.5671	.0732	.7550	43
18	.36325	.93169	.38988	.5649	.0733	.7529	42
19	.36352	.93158	.39022	.5627	.0734	.7509	41
20	.36379	.93148	.39055	2.5605	1.0736	2.7488	40
21	.36406	.93137	.39089	.5583	.0737	.7468	39
22	.36433	.93127	.39122	.5561	.0738	.7447	38
23	.36460	.93116	.39156	.5539	.0739	.7427	37
24	.36488	.93105	.39189	.5517	.0740	.7406	36
25	.36515	.93095	.39223	2.5495	1.0742	2.7386	35
26	.36542	.93084	.39257	.5473	.0743	.7366	34
27	.36569	.93074	.39290	.5451	.0744	.7346	33
28	.36596	.93063	.39324	.5430	.0745	.7325	32
29	.36623	.93052	.39357	.5408	.0747	.7305	31
30	.36650	.93042	.39391	2.5386	1.0748	2.7285	30
31	.36677	.93031	.39425	.5365	.0749	.7265	29
32	.36704	.93020	.39458	.5343	.0750	.7245	28
33	.36731	.93010	.39492	.5322	.0751	.7225	27
34	.36758	.92999	.39525	.5300	.0753	.7205	26
35	.36785	.92988	.39559	2.5278	1.0754	2.7185	25
36	.36812	.92978	.39593	.5257	.0755	.7165	24
37	.36839	.92967	.39626	.5236	.0756	.7145	23
38	.36866	.92956	.39660	.5214	.0758	.7125	22
39	.36893	.92945	.39694	.5193	.0759	.7105	21
40	.36921	.92935	.39727	2.5171	1.0760	2.7085	20
41	.36948	.92924	.39761	.5150	.0761	.7065	19
42	.36975	.92913	.39795	.5129	.0763	.7045	18
43	.37002	.92902	.39828	.5108	.0764	.7026	17
44	.37029	.92892	.39862	.5086	.0765	.7006	16
45	.37056	.92881	.39896	2.5065	1.0766	2 6986	15
46	.37083	.92870	.39930	.5044	.0768	.6967	14
47	.37110	.92859	.39963	.5023	.0769	.6947	13
48	.37137	.92848	.39997	.5002	.0770	.6927	12
49	.37164	.92838	.40031	.4981	.0771	.6908	11
50	.37191	.92827	.40065	2.4960	1.0773	2.6888	10
51	.37218	.92816	.40098	.4939	.0774	.6869	9
52	.37245	.92805	.40132	.4918	.0775	.6849	8
53	.37272	.92794	.40166	.4897	.0776	.6830	7
54	.37299	.92784	.40200	.4876	.0778	.6810	6
55	.37326	.92773	.40233	2.4855	1.0779	2.6791	5
56	.37353	.92762	.40267	.4834	.0780	.6772	4
57	.37380	.92751	.40301	.4813	.0781	.6752	3
58	.37407	.97740	.40335	.4792	.0783	.6733	2
59	.37434	.92729	.40369	.4772	.0784	.6714	1
60	.37461	.92718	.40403	2.4751	1.0785	2.6695	0
M	Cosine	Sine	Cotan.	Tan.	Cosec.	Secant	M

68°

22°

M	Sine	Cosine	Tan.	Cotan.	Secant	Cosec.	M
0	.37461	.92718	.40403	2.4751	1.0785	2.6695	60
1	.37488	.92707	.40436	.4730	.0787	.6675	59
2	.37514	.92696	.40470	.4709	.0788	.6656	58
3	.37541	.92686	.40504	.4689	.0789	.6637	57
4	.37568	.92675	.40538	.4668	.0790	.6618	56
5	.37595	.92664	.40572	2.4647	1.0792	2.6599	55
6	.37622	.92653	.40606	.4627	.0793	.6580	54
7	.37649	.92642	.40640	.4606	.0794	.6561	53
8	.37676	.92631	.40673	.4586	.0795	.6542	52
9	.37703	.92620	.40707	.4565	.0797	.6523	51
10	.37730	.92609	.40741	2.4545	1.0798	2.6504	50
11	.37757	.92598	.40775	.4525	.0799	.6485	49
12	.37784	.92587	.40809	.4504	.0801	.6466	48
13	.37811	.92576	.40843	.4484	.0802	.6447	47
14	.37838	.92565	.40877	.4463	.0803	.6428	46
15	.37865	.92554	.40911	2.4443	1.0804	2 6410	45
16	.37892	.92543	.40945	.4423	.0806	.6391	44
17	.37919	.92532	.40979	.4403	.0807	.6372	43
18	.37946	.92521	.41013	.4382	.0808	.6353	42
19	.37972	.92510	.41047	.4362	.0610	.6335	41
20	.37999	.92499	.41081	2.4342	1.0811	2.6316	40
21	.38026	.92488	.41115	.4322	.0812	.6297	39
22	.38053	.92477	.41149	.4302	.0813	.6279	38
23	.38080	.92466	.41183	.4282	.0815	.6260	37
24	.38107	.92455	.41217	.4262	.0816	.6242	36
25	.38134	.92443	.41251	2.4242	1.0817	2.6223	35
26	.38161	.92432	.41285	.4222	.0819	.6205	34
27	.38188	.92421	.41319	.4202	.0820	.6186	33
28	.38214	.92410	.41353	.4182	.0821	.6168	32
29	.38241	.92399	.41387	.4162	.0823	.6150	31
30	.38268	.92388	.41421	2.4142	1.0824	2.6131	30
31	.38295	.92377	.41455	.4122	.0825	.6113	29
32	.38322	.92366	.41489	.4102	.0826	.6095	28
33	.38349	.92354	.41524	.4083	.0828	.6076	27
34	.38376	.92343	.41558	.4063	.0829	.6058	26
35	.38403	.92332	.41592	2.4043	1.0830	2.6040	25
36	.38429	.92321	.41626	.4023	.0832	.6022	24
37	.38456	.92310	.41660	.4004	.0833	.6003	23
38	.38483	.92299	.41694	.3984	.0834	.5985	22
39	.38510	.92287	.41728	.3964	.0836	.5967	21
40	.38537	.92276	.41762	2.3945	1.0837	2.5949	20
41	.38564	.92265	.41797	.3925	.0838	.5931	19
42	.38591	.92254	.41831	.3906	.0840	.5913	18
43	.38617	.92242	.41865	.3886	.0841	.5895	17
44	.38644	.92231	.41899	.3867	.0842	.5877	16
45	.38671	.92220	.41933	2.3847	1.0844	2.5859	15
46	.38698	.92209	.41968	.3828	.0845	.5841	14
47	.38725	.92197	.42002	.3808	.0846	.5823	13
48	.38751	.92186	.42036	.3789	.0847	.5805	12
49	.38778	.92175	.42070	.3770	.0849	.5787	11
50	.38805	.92164	.42105	2.3750	1.0850	2.5770	10
51	.38832	.92152	.42139	.3731	.0851	.5752	9
52	.38859	.92141	.42173	.3712	.0853	.5734	8
53	.38886	.92130	.42207	.3692	.0854	.5716	7
54	.38912	.92118	.42742	.3673	.0855	.5699	6
55	.38939	.92107	.42276	2 3654	1.0857	7.5681	5
56	.38966	.92096	.42310	.3635	.0858	.5663	4
57	.38993	.92084	.42344	.3616	.0859	.5646	3
58	.39019	.92073	.42379	.3597	.0861	.5628	2
59	.39046	.92062	.42413	.3577	.0862	.5610	1
60	.39073	.92050	.42447	2.3558	1.0864	2.5593	0
M	Cosine	Sine	Cotan.	Tan.	Cosec.	Secant	M

67°

23°

M	Sine	Cosine	Tan.	Cotan.	Secant	Cosec.	M
0	.39073	.92050	.42447	2.3558	1.0864	2.5593	60
1	.39100	.92039	.42482	.3539	.0865	.5575	59
2	.39126	.92028	.42516	.3520	.0866	.5558	58
3	.39153	.92016	.42550	.3501	.0868	.5540	57
4	.39180	.92005	.42585	.3482	.0869	.5523	56
5	.39207	.91993	.42619	2.3463	1.0870	2.5506	55
6	.39234	.91982	.42654	.3445	.0872	.5488	54
7	.39260	.91971	.42688	.3426	.0873	.5471	53
8	.39287	.91959	.42722	.3407	.0874	.5453	52
9	.39314	.91948	.42757	.3388	.0876	.5436	51
10	.39341	.91936	.42791	2.3369	1.0877	2.5419	50
11	.39367	.91925	.42826	.3350	.0878	.5402	49
12	.39394	.91913	.42860	.3332	.0880	.5384	48
13	.39421	.91902	.42894	.3313	.0881	.5367	47
14	.39448	.91891	.42929	.3294	.0882	.5350	46
15	.39474	.91879	.42963	2.3276	1.0884	2.5333	45
16	.39501	.91868	.42998	.3257	.0885	.5316	44
17	.39528	.91856	.43032	.3238	.0886	.5299	43
18	.39554	.91845	.43067	.3220	.0888	.5281	42
19	.39581	.91833	.43101	.3201	.0889	.5264	41
20	.39608	.91822	.43136	2.3183	1.0891	2.5247	40
21	.39635	.91810	.43170	.3164	.0892	.5230	39
22	.39661	.91798	.43205	.3145	.0893	.5213	38
23	.39688	.91787	.43239	.3127	.0895	.5196	37
24	.39715	.91775	.43274	.3109	.0896	.5179	36
25	.39741	.91764	.43308	2.3090	1.0897	2.5163	35
26	.39768	.91752	.43343	.3072	.0899	.5146	34
27	.39795	.91741	.43377	.3053	.0900	.5129	33
28	.39821	.91729	.43412	.3035	.0902	.5112	32
29	.39848	.91718	.43447	.3017	.0903	.5095	31
30	.39875	.91706	.43481	2.2998	1.0904	2.5078	30
31	.39901	.91694	.43516	.2980	.0906	.5062	29
32	.39928	.91683	.43550	.2962	.0907	.5045	28
33	.39955	.91671	.43585	.2944	.0908	.5028	27
34	.39981	.91659	.43620	.2925	.0910	.5011	26
35	.40008	.91648	.43654	2.2907	1.0911	2.4995	25
36	.40035	.91636	.43689	.2889	.0913	.4978	24
37	.40061	.91625	.43723	.2871	.0914	.4961	23
38	.40088	.91613	.43758	.2853	.0915	.4945	22
39	.40115	.91601	.43793	.2835	.0917	.4928	21
40	.40141	.91590	.43827	2.2817	1.0918	2.4912	20
41	.40168	.91578	.43862	.2799	.0920	.4895	19
42	.40195	.91566	.43897	.2781	.0921	.4879	18
43	.40221	.91554	.43932	.2763	.0922	.4862	17
44	.40248	.91543	.43966	.2745	.0924	.4846	16
45	.40275	.91531	.44001	2.2727	1.0925	2.4829	15
46	.40301	.91519	.44036	.2709	.0927	.4813	14
47	.40328	.91508	.44070	.2691	.0928	.4797	13
48	.40354	.91496	.44105	.2673	.0929	.4780	12
49	.40381	.91484	.44140	.2655	.0931	.4764	11
50	.40408	.91472	.44175	2.2637	1.0932	2.4748	10
51	.40434	.91461	.44209	.2619	.0934	.4731	9
52	.40461	.91449	.44244	.2602	.0935	.4715	8
53	.40487	.91437	.44279	.2584	.0936	.4699	7
54	.40514	.91425	.44314	.2566	.0938	.4683	6
55	.40541	.91414	.44349	2.2548	1.0939	2.4666	5
56	.40567	.91402	.44383	.2531	.0941	.4650	4
57	.40594	.91390	.44418	.2513	.0942	.4634	3
58	.40620	.91378	.44453	.2495	.0943	.4618	2
59	.40647	.91366	.44488	.2478	.0945	.4602	1
60	.40674	.91354	.44523	2.2460	1.0946	2.4586	0
M	Cosine	Sine	Cotan.	Tan.	Cosec.	Secant	M

66°

24°

M	Sine	Cosine	Tan.	Cotan.	Secant	Cosec.	M
0	.40674	.91354	.44523	2.2460	1.0946	2.4586	60
1	.40700	.91343	.44558	.2443	.0948	.4570	59
2	.40727	.91331	.44593	.2425	.0949	.4554	58
3	.40753	.91319	.44627	.2408	.0951	.4538	57
4	.40780	.91307	.44662	.2390	.0952	.4522	56
5	.40806	.91295	.44697	2.2373	1.0953	2.4506	55
6	.40833	.91283	.44732	.2355	.0955	.4490	54
7	.40860	.91271	.44767	.2338	.0956	.4474	53
8	.40886	.91260	.44802	.2320	.0958	.4458	52
9	.40913	.91248	.44837	.2303	.0959	.4442	51
10	.40939	.91236	.44872	2.2286	1.0961	2.4426	50
11	.40966	.91224	.44907	.2268	.0962	.4411	49
12	.40992	.91212	.44942	.2251	.0963	.4395	48
13	.41019	.91200	.44977	.2234	.0965	.4379	47
14	.41045	.91188	.45012	.2216	.0966	.4363	46
15	.41072	.91176	.45047	2.2199	1.0968	2.4347	45
16	.41098	.91164	.45082	.2182	.0969	.4332	44
17	.41125	.91152	.45117	.2165	.0971	.4316	43
18	.41151	.91140	.45152	.2147	.0972	.4300	42
19	.41178	.91128	.45187	.2130	.0973	.4285	41
20	.41204	.91116	.45222	2.2113	1.0975	2.4269	40
21	.41231	.91104	.45257	.2096	.0976	.4254	39
22	.41257	.91092	.45292	.2079	.0978	.4238	38
23	.41284	.91080	.45327	.2062	.0979	.4222	37
24	.41310	.91068	.45362	.2045	.0981	.4207	36
25	.41337	.91056	.45397	2.2028	1.0982	2.4191	35
26	.41363	.91044	.45432	.2011	.0984	.4176	34
27	.41390	.91032	.45467	.1994	.0985	.4160	33
28	.41416	.91020	.45502	.1977	.0986	.4145	32
29	.41443	.91008	.45537	.1960	.0988	.4130	31
30	.41469	.90996	.45573	2.1943	1.0989	2.4114	30
31	.41496	.90984	.45608	.1926	.0991	.4099	29
32	.41522	.90972	.45643	.1909	.0992	.4083	28
33	.41549	.90960	.45678	.1892	.0994	.4068	27
34	.41575	.90948	.45713	.1875	.0995	.4053	26
35	.41602	.90936	.45748	2.1859	1.0997	2.4037	25
36	.41628	.90924	.45783	.1842	.0998	.4022	24
37	.41654	.90911	.45819	.1825	.1000	.4007	23
38	.41681	.90899	.45854	.1808	.1001	.3992	22
39	.41707	.90887	.45889	.1792	.1003	.3976	21
40	.41734	.90875	.45924	2.1775	1.1004	2.3961	20
41	.41760	.90863	.45960	.1758	.1005	.3946	19
42	.41787	.90851	.45995	.1741	.1007	.3931	18
43	.41813	.90839	.46030	.1725	.1008	.3916	17
44	.41839	.90826	.46065	.1708	.1010	.3901	16
45	.41866	.90814	.46101	2.1692	1.1011	2.3886	15
46	.41892	.90802	.46136	.1675	.1013	.3871	14
47	.41919	.90790	.46171	.1658	.1014	.3856	13
48	.41945	.90778	.46206	.1642	.1016	.3841	12
49	.41972	.90765	.46242	.1625	.1017	.3826	11
50	.41998	.90753	.46277	2.1609	1.1019	2.3811	10
51	.42024	.90741	.46312	.1592	.1020	.3796	9
52	.42051	.90729	.46348	.1576	.1022	.3781	8
53	.42077	.90717	.46383	.1559	.1023	.3766	7
54	.42103	.90704	.46418	.1543	.1025	.3751	6
55	.42130	.90692	.46454	2.1527	1.1026	2.3736	5
56	.42156	.90680	.46489	.1510	.1028	.3721	4
57	.42183	.90668	.46524	.1494	.1029	.3706	3
58	.42209	.90655	.46560	.1478	.1031	.3691	2
59	.42235	.90643	.46595	.1461	.1032	.3677	1
60	.42262	.90631	.46631	2.1445	1.1034	2.3662	0
M	Cosine	Sine	Cotan.	Tan.	Cosec.	Secant	M

65°

25°

M	Sine	Cosine	Tan.	Cotan.	Secant	Cosec.	M
0	.42262	.90631	.46631	2.1445	1.1034	2.3662	60
1	.42288	.90618	.46666	.1429	.1035	.3647	59
2	.42314	.90606	.46702	.1412	.1037	.3632	58
3	.42341	.90594	.46737	.1396	.1038	.3618	57
4	.42367	.90581	.46772	.1380	.1040	.3603	56
5	.42394	.90569	.46808	2.1364	1.1041	2.3588	55
6	.42420	.90557	.46843	.1348	.1043	.3574	54
7	.42446	.90544	.46879	.1331	.1044	.3559	53
8	.42473	.90532	.46914	.1315	.1046	.3544	52
9	.42499	.90520	.46950	.1299	.1047	.3530	51
10	.42525	.90507	.46985	2.1283	1.1049	2.3515	50
11	.42552	.90495	.47021	.1267	.1050	.3501	49
12	.42578	.90483	.47056	.1251	.1052	.3486	48
13	.42604	.90470	.47092	.1235	.1053	.3472	47
14	.42630	.90458	.47127	.1219	.1055	.3457	46
15	.42657	.90445	.47163	2.1203	1.1056	2.3443	45
16	.42683	.90433	.47199	.1187	.1058	.3428	44
17	.42709	.90421	.47234	.1171	.1059	.3414	43
18	.42736	.90408	.47270	.1155	.1061	.3399	42
19	.42762	.90396	.47305	.1139	.1062	.3385	41
20	.42788	.90383	.47341	2.1123	1.1064	2.3371	40
21	.42815	.90371	.47376	.1107	.1065	.3356	39
22	.42841	.90358	.47412	.1092	.1067	.3342	38
23	.42867	.90346	.47448	.1076	.1068	.3328	37
24	.42893	.90333	.47483	.1060	.1070	.3313	36
25	.42920	.90321	.47519	2.1044	1.1072	2.3299	35
26	.42946	.90308	.47555	.1028	.1073	.3285	34
27	.42972	.90296	.47590	.1013	.1075	.3271	33
28	.42998	.90283	.47626	.0997	.1076	.3256	32
29	.43025	.90271	.47662	.0981	.1078	.3242	31
30	.43051	.90258	.47697	2.0965	1.1079	2.3228	30
31	.43077	.90246	.47733	.0950	.1081	.3214	29
32	.43104	.90233	.47769	.0934	.1082	.3200	28
33	.43130	.90221	.47805	.0918	.1084	.3186	27
34	.43156	.90208	.47840	.0903	.1085	.3172	26
35	.43182	.90196	.47876	2.0887	1.1087	2.3158	25
36	.43208	.90183	.47912	.0872	.1088	.3143	24
37	.43235	.90171	.47948	.0856	.1090	.3129	23
38	.43261	.90158	.47983	.0840	.1092	.3115	22
39	.43287	.90145	.48019	.0825	.1093	.3101	21
40	.43313	.90133	.48055	2.0809	1.1095	2.3087	20
41	.43340	.90120	.48091	.0794	.1096	.3073	19
42	.43366	.90108	.48127	.0778	.1098	.3059	18
43	.43392	.90095	.48162	.0763	.1099	.3046	17
44	.43418	.90082	.48198	.0747	.1101	.3032	16
45	.43444	.90070	.48234	2.0732	1.1102	2.3018	15
46	.43471	.90057	.48270	.0717	.1104	.3004	14
47	.43497	.90044	.48306	.0701	.1106	.2990	13
48	.43523	.90032	.48342	.0686	.1107	.2976	12
49	.43549	.90019	.48378	.0671	.1109	.2962	11
50	.43575	.90006	.48414	2.0655	1.1110	2.2949	10
51	.43602	.89994	.48449	.0640	.1112	.2935	9
52	.43628	.89981	.48485	.0625	.1113	.2921	8
53	.43654	.89968	.48521	.0609	.1115	.2907	7
54	.43680	.89956	.48557	.0594	.1116	.2894	6
55	.43706	.89943	.48593	2.0579	1.1118	2.2880	5
56	.43732	.89930	.48629	.0564	.1120	.2866	4
57	.43759	.89918	.48665	.0548	.1121	.2853	3
58	.43785	.89905	.48701	.0533	.1123	.2839	2
59	.43811	.89892	.48737	.0518	.1124	.2825	1
60	.43837	.89879	.48773	2.0503	1.1126	2.2812	0
M	Cosine	Sine	Cotan.	Tan.	Cosec.	Secant	M

64°

26°

M	Sine	Cosine	Tan.	Cotan.	Secant	Cosec.	M
0	.43837	.89879	.48773	2.0503	1.1126	2.2812	60
1	.43863	.89867	.48809	.0488	.1127	.2798	59
2	.43889	.89854	.48845	.0473	.1129	.2784	58
3	.43915	.89841	.48881	.0458	.1131	.2771	57
4	.43942	.89828	.48917	.0443	.1132	.2757	56
5	.43968	.89815	.48953	2.0427	1.1134	2.2744	55
6	.43994	.89803	.48989	.0412	.1135	.2730	54
7	.44020	.89790	.49025	.0397	.1137	.2717	53
8	.44046	.89777	.49062	2.0382	.1139	.2703	52
9	.44072	.89764	.49098	.0367	.1140	.2690	51
10	.44098	.89751	.49134	2.0352	1.1142	2.2676	50
11	.44124	.89739	.49170	.0338	.1143	.2663	49
12	.44150	.89726	.49206	.0323	.1145	.2650	48
13	.44177	.89713	.49242	.0308	.1147	.2636	47
14	.44203	.89700	.49278	.0293	.1148	.2623	46
15	.44229	.89687	.49314	2.0278	1.1150	2.2610	45
16	.44255	.89674	.49351	.0263	.1151	.2596	44
17	.44281	.89661	.49387	.0248	.1153	.2583	43
18	.44307	.89649	.49423	.0233	.1155	.2570	42
19	.44333	.89636	.49459	.0219	.1156	.2556	41
20	.44359	.89623	.49495	2.0204	1.1158	2.2543	40
21	.44385	.89610	.49532	.0189	.1159	.2530	39
22	.44411	.89597	.49568	.0174	.1161	.2517	38
23	.44437	.89584	.49604	.0159	.1163	.2503	37
24	.44463	.89571	.49640	.0145	.1164	.2490	36
25	.44489	.89558	.49677	2.0130	1.1166	2.2477	35
26	.44516	.89545	.49713	.0115	.1167	.2464	34
27	.44542	.89532	.49749	.0101	.1169	.2451	33
28	.44568	.89519	.49785	.0086	.1171	.2438	32
29	.44594	.89506	.49822	.0071	.1172	.2425	31
30	.44620	.89493	.49858	2.0057	1.1174	2.2411	30
31	.44646	.89480	.49894	.0042	.1176	.2398	29
32	.44672	.89467	.49931	.0028	.1177	.2385	28
33	.44698	.89454	.49967	.0013	.1179	.2372	27
34	.44724	.89441	.50003	1.9998	.1180	.2359	26
35	.44750	.89428	.50040	1.9984	1.1182	2.2348	25
36	.44776	.89415	.50076	.9969	.1184	.2333	24
37	.44802	.89402	.50113	.9955	.1185	.2320	23
38	.44828	.89389	.50149	.9940	.1187	.2307	22
39	.44854	.89376	.50185	.9926	.1189	.2294	21
40	.44880	.89363	.50222	1.9912	1.1190	2.2282	20
41	.44906	.89350	.50258	.9897	.1192	.2269	19
42	.44932	.89337	.50295	.9883	.1193	.2256	18
43	.44958	.89324	.50331	.9868	.1195	.2243	17
44	.44984	.89311	.50368	.9854	.1197	.2230	16
45	.45010	.89298	.50404	1.9840	1.1198	2.2217	15
46	.45036	.89285	.50441	.9825	.1200	.2204	14
47	.45062	.89272	.50477	.9811	.1202	.2192	13
48	.45088	.89258	.50514	.9797	.1203	.2179	12
49	.45114	.89245	.50550	.9782	.1205	.2166	11
50	.45140	.89232	.50587	1.9768	1.1207	2.2153	10
51	.45166	.89219	.50623	.9754	.1208	.2141	9
52	.45191	.89206	.50660	.9739	.1210	.2128	8
53	.45217	.89193	.50696	.9725	.1212	.2115	7
54	.45243	.89180	.50733	.9711	.1213	.2103	6
55	.45269	.89166	.50769	1.9697	1.1215	2.2090	5
56	.45295	.89153	.50806	.9683	.1217	.2077	4
57	.45321	.89140	.50843	.9668	.1218	.2065	3
58	.45347	.89127	.50879	.9654	.1220	.2052	2
59	.45373	.89114	.50916	.9640	.1222	.2039	1
60	.45399	.89101	.50952	1.9626	1.1223	2.2027	0
M	Cosine	Sine	Cotan.	Tan.	Cosec.	Secant	M

63°

27°

M	Sine	Cosine	Tan.	Cotan.	Secant	Cosec.	M
0	.45399	.89101	.50952	1.9626	1.1223	2.2027	60
1	.45425	.89087	.50989	.9612	.1225	.2014	59
2	.45451	.89074	.51026	.9598	.1226	.2002	58
3	.45477	.89061	.51062	.9584	.1228	.1989	57
4	.45503	.89048	.51099	.9570	.1230	.1977	56
5	.45528	.89034	.51136	1.9556	1.1231	2.1964	55
6	.45554	.89021	.51172	.9542	.1233	.1952	54
7	.45580	.89008	.51209	.9528	.1235	.1939	53
8	.45606	.88995	.51246	.9514	.1237	.1927	52
9	.45632	.88981	.51283	.9500	.1238	.1914	51
10	.45658	.88968	.51319	1.9486	1.1240	2.1902	50
11	.45684	.88955	.51356	.9472	.1242	.1889	49
12	.45710	.88942	.51393	.9458	.1243	.1877	48
13	.45736	.88928	.51430	.9444	.1245	.1865	47
14	.45761	.88915	.51466	.9430	.1247	.1852	46
15	.45787	.88902	.51503	1.9416	1.1248	2.1840	45
16	.45813	.88888	.51540	.9402	.1250	.1828	44
17	.45839	.88875	.51577	.9388	.1252	.1815	43
18	.45865	.88862	.51614	.9375	.1253	.1803	42
19	.45891	.88848	.51651	.9361	.1255	.1791	41
20	.45917	.88835	.51687	1.9347	1.1257	2.1778	40
21	.45942	.88822	.51724	.9333	.1258	.1766	39
22	.45968	.88808	.51761	.9319	.1260	.1754	38
23	.45994	.88795	.51798	.9306	.1262	.1742	37
24	.46020	.88781	.51835	.9292	.1264	.1730	36
25	.46046	.88768	.51872	1.9278	1.1265	2.1717	35
26	.46072	.88755	.51909	.9264	.1267	.1705	34
27	.46097	.88741	.51946	.9251	.1269	.1693	33
28	.46123	.88728	.51983	.9237	.1270	.1681	32
29	.46149	.88714	.52020	.9223	.1272	.1669	31
30	46175	.88701	.52057	1.9210	1.1274	2.1657	30
31	.46201	.88688	.52094	.9196	.1275	.1645	29
32	.46226	.88674	.52131	.9182	.1277	.1633	28
33	.46252	.88661	.52168	.9169	.1279	.1620	27
34	.46278	.88647	.52205	.9155	.1281	.1608	26
35	.46304	.88634	.52242	1.9142	1.1282	2.1596	25
36	.46330	.88620	.52279	.9128	.1284	.1584	24
37	.46355	.88607	.52316	.9115	.1286	.1572	23
38	.46381	.88593	.52353	.9101	.1287	.1560	22
39	.46407	.88580	.52390	.9088	.1289	.1548	21
40	.46433	.88566	.52427	1.9074	1.1291	2.1536	20
41	.46458	.88553	.52464	.9061	.1293	.1525	19
42	.46484	.88539	.52501	.9047	.1294	.1513	18
43	.46510	.88526	.52538	.9034	.1296	.1501	17
44	.46536	.88512	.52575	.9020	.1298	.1489	16
45	.46561	.88499	.52612	1.9007	1.1299	2.1477	15
46	.46587	.88485	.52650	.8993	.1301	.1465	14
47	.46613	.88472	.52687	.8980	.1303	.1453	13
48	.46639	.88458	.52724	.8967	.1305	.1441	12
49	.46664	.88444	.52761	.8953	.1306	.1430	11
50	.46690	.88431	.52798	1.8940	1.1308	2.1418	10
51	.46716	.88417	.52836	.8927	.1310	.1406	9
52	.46741	.88404	.52873	.8913	.1312	.1394	8
53	.46767	.88390	.52910	.8900	.1313	.1382	7
54	.46793	.88376	.52947	.8887	.1315	.1371	6
55	.46819	.88363	.52984	1.8873	1.1317	2.1359	5
56	.46844	.88349	.53022	.8860	.1319	.1347	4
57	.46870	.88336	.53059	.8847	.1320	.1335	3
58	.46896	.88322	53096	.8834	.1322	.1324	2
59	.46921	.88308	.53134	.8820	.1324	.1312	1
60	.46947	.88295	.53171	1.8807	1.1326	2.1300	0
M	Cosine	Sine	Cotan.	Tan.	Cosec.	Secant	M

62°

28°

M	Sine	Cosine	Tan.	Cotan.	Secant	Cosec.	M
0	.46947	.88295	.53171	1.8807	1.1326	2.1300	60
1	.46973	.88281	.53208	.8794	.1327	.1289	59
2	.46998	.88267	.53245	.8781	.1329	.1277	58
3	.47024	.88254	.53283	.8768	.1331	.1266	57
4	.47050	.88240	.53320	.8754	.1333	.1254	56
5	.47075	.88226	.53358	1.8741	1.1334	2.1242	55
6	.47101	.88213	.53395	.8728	.1336	.1231	54
7	.47127	.88199	.53432	.8715	.1338	.1219	53
8	.47152	.88185	.53470	.8702	.1340	.1208	52
9	.47178	.88171	.53507	.8689	.1341	.1196	51
10	.47204	.88158	.53545	1.8676	1.1343	2.1185	50
11	.47229	.88144	.53582	.8663	.1345	.1173	49
12	.47255	.88130	.53619	.8650	.1347	.1162	48
13	.47281	.88117	.53657	.8637	.1349	.1150	47
14	.47306	.88103	.53694	.8624	.1350	.1139	46
15	.47332	.88089	.53732	1.8611	1.1352	2.1127	45
16	.47357	.88075	.53769	.8598	.1354	.1116	44
17	.47383	.88061	.53807	.8585	.1356	.1104	43
18	.47409	.88048	.53844	.8572	.1357	.1093	42
19	.47434	.88034	.53882	.8559	.1359	.1082	41
20	.47460	.88020	.53919	1.8546	1.1361	2.1070	40
21	.47486	.88006	.53957	.8533	.1363	.1059	39
22	.47511	.87992	.53995	.8520	.1365	.1048	38
23	.47537	.87979	.54032	.8507	.1366	.1036	37
24	.47562	.87965	.54070	.8495	.1368	.1025	36
25	.47588	.87951	.54107	1.8482	1.1370	2.1014	35
26	.47613	.87937	.54145	.8469	.1372	.1002	34
27	.47639	.87923	.54183	.8456	.1373	.0991	33
28	.47665	.87909	.54220	.8443	.1375	.0980	32
29	.47690	.87895	.54258	.8430	.1377	.0969	31
30	.47716	.87882	.54295	1.8418	1.1379	2.0957	30
31	.47741	.87868	.54333	.8405	.1381	.0946	29
32	.47767	.87854	.54371	.8392	.1382	.0935	28
33	.47792	.87840	.54409	.8379	.1384	.0924	27
34	.47818	.87826	.54446	.8367	.1386	.0912	26
35	.47844	.87812	.54484	1.8354	1.1388	2.0901	25
36	.47869	.87798	.54522	.8341	.1390	.0890	24
37	.47895	.87784	.54559	.8329	.1391	.0879	23
38	.47920	.87770	.54597	.8316	.1393	.0868	22
39	.47946	.87756	.54635	.8303	.1395	.0857	21
40	.47971	.87742	.54673	1.8291	1.1397	2.0846	20
41	.47997	.87728	.54711	.8278	.1399	.0835	19
42	.48022	.87715	.54748	.8265	.1401	.0824	18
43	.48048	.87701	.54786	.8253	.1402	.0812	17
44	.48073	.87687	.54824	.8240	.1404	.0801	16
45	.48099	.87673	.54862	1.8227	1.1406	2.0790	15
46	.48124	.87659	.54900	.8215	.1408	.0779	14
47	.48150	.87645	.54937	.8202	.1410	.0768	13
48	.48175	.87631	.54975	.8190	.1411	.0757	12
49	.48201	.87617	.55013	.8177	.1413	.0746	11
50	.48226	.87603	.55051	1.8165	1.1415	2.0735	10
51	.48252	.87588	.55089	.8152	.1417	.0725	9
52	.48277	.87574	.55127	.8140	.1419	.0714	8
53	.48303	.87560	.55165	.8127	.1421	.0703	7
54	.48328	.87546	.55203	.8115	.1422	.0692	6
55	.48354	.87532	.55241	1.8102	1.1424	2.0681	5
56	.48379	.87518	.55279	.8090	.1426	.0670	4
57	.48405	.87504	.55317	.8078	.1428	.0659	3
58	.48430	.87490	.55355	.8065	.1430	.0648	2
59	.48455	.87476	.55393	.8053	.1432	.0637	1
60	.48481	.87462	.55431	1.8040	1.1433	2.0627	0
M	Cosine	Sine	Cotan.	Tan.	Cosec.	Secant	M

61°

29°

M	Sine	Cosine	Tan.	Cotan.	Secant	Cosec.	M
0	.48481	.87462	.55431	1.8040	1.1433	2.0627	60
1	.48506	.87448	.55469	.8028	.1435	.0616	59
2	.48532	.87434	.55507	.8016	.1437	.0605	58
3	.48557	.87420	.55545	.8003	.1439	.0594	57
4	.48583	.87405	.55583	.7991	.1441	.0583	56
5	.48608	.87391	.55621	1.7979	1.1443	2.0573	55
6	.48633	.87377	.55659	.7966	.1445	.0562	54
7	.48659	.87363	.55697	.7954	.1446	0551	53
8	.48684	.87349	.55735	.7942	.1448	.0540	52
9	.48710	.87335	.55774	.7930	.1450	.0530	51
10	.48735	.87320	.55812	1.7917	1.1452	2.0519	50
11	.48760	.87306	.55850	.7905	.1454	.0508	49
12	.48786	.87292	.55888	.7893	.1456	.0498	48
13	.48811	.87278	.55926	.7881	.1458	.0487	47
14	.48837	.87264	.55964	.7868	.1459	.0476	46
15	.48862	.87250	.56003	1.7856	1.1461	2.0466	45
16	.48887	.87235	.56041	.7844	.1463	.0455	44
17	.48913	.87221	.56079	.7832	.1465	.0444	43
18	.48938	.87207	.56117	.7820	.1467	.0434	42
19	.48964	.87193	.56156	.7808	.1469	.0423	41
20	.48989	.87178	.56194	1.7795	1.1471	2.0413	40
21	.49014	.87164	.56232	.7783	.1473	.0402	39
22	.49040	.87150	.56270	.7771	.1474	.0392	38
23	.49065	.87136	.56309	.7759	.1476	.0381	37
24	.49090	.87121	.56347	.7747	.1478	.0370	36
25	.49116	.87107	.56385	1.7735	1.1480	2.0360	35
26	.49141	.87093	.56424	.7723	.1482	.0349	34
27	.49166	.87078	.56462	.7711	.1484	.0339	33
28	.49192	.87064	.56500	.7699	.1486	.0329	32
29	.49217	.87050	.56539	.7687	.1488	.0318	31
30	.49242	.87035	.56577	1.7675	1.1489	2.0308	30
31	.49268	.87021	.56616	.7663	.1491	.0297	29
32	.49293	.87007	.56654	.7651	.1493	.0287	28
33	.49318	86992	.56692	.7639	.1495	.0276	27
34	.49343	.86978	.56731	.7627	.1497	.0266	26
35	.49369	.86964	.56769	1.7615	1.1499	2.0256	25
36	.49394	.86949	.56808	.7603	.1501	.0245	24
37	.49419	.86935	.56846	.7591	.1503	.0235	23
38	.49445	.86921	.56885	.7579	.1505	.0224	22
39	.49470	.86906	.56923	.7567	.1507	.0214	21
40	.49495	.86892	.56962	1.7555	1.1508	2.0204	20
41	.49521	.86877	.57000	.7544	.1510	.0194	19
42	.49546	.86863	.57039	.7532	.1512	.0183	18
43	.49571	.86849	.57077	.7520	.1514	.0173	17
44	.49596	.86834	.57116	.7508	.1516	.0163	16
45	.49622	.86820	.57155	1.7496	1.1518	2.0152	15
46	.49647	.86805	.57193	.7484	.1520	.0142	14
47	.49672	.86791	.57232	.7473	.1522	.0132	13
48	.49697	.86776	.57270	.7461	.1524	.0122	12
49	.49723	.86762	.57309	.7449	.1526	.0111	11
50	.49748	.86748	.57348	1.7437	1.1528	2.0101	10
51	.49773	.86733	.57386	.7426	.1530	.0091	9
52	.49798	.86719	.57425	.7414	.1531	.0081	8
53	.49823	.86704	.57464	.7402	.1533	.0071	7
54	.49849	.86690	.57502	.7390	.1535	.0061	6
55	.49874	.86675	.57541	1.7379	1.1537	2.0050	5
56	.49899	.86661	.57580	.7367	.1539	.0040	4
57	.49924	.86646	.57619	.7355	.1541	.0030	3
58	.49950	.86632	.57657	.7344	.1543	.0020	2
59	.49975	.86617	.57696	.7332	.1545	.0010	1
60	.50000	.86603	.57735	1.7320	1.1547	2.0000	0
M	Cosine	Sine	Cotan.	Tan.	Cosec.	Secant	M

60°

30°

M	Sine	Cosine	Tan.	Cotan.	Secant	Cosbc.	M
0	.50000	.86603	.57735	1.7320	1.1547	2.0000	60
1	.50025	.86588	.67774	.7309	.1549	1.9990	59
2	.50050	.86573	.57813	.7297	.1551	.9980	58
3	.50075	86559	.57851	.7286	.1553	.9970	57
4	.50101	86544	.57890	.7274	.1555	9960	56
5	.50126	.86530	.57929	1.7262	1.1557	1.9950	55
6	.50151	86515	.57968	.7251	.1559	.9940	54
7	.50176	.86500	.58007	.7239	.1561	.9930	53
8	.50201	86486	.58046	.7228	1562	.9920	52
9	.50226	.86471	.58085	.7216	1564	.9910	51
10	.50252	.86457	.58123	1.7205	1.1566	1.9900	50
11	.50277	.86442	.58162	.7193	.1568	.9890	49
12	.50302	.86427	.58201	.7182	.1570	9880	48
13	.50327	.86413	.58240	.7170	1572	.9870	47
14	.50352	.86398	.58279	.7159	.1574	.9860	46
15	.50377	.86383	.58318	1.7147	1.1576	1.9850	45
16	.50402	.86369	.58357	.7136	.1578	.9840	44
17	.50428	.86354	.58396	.7124	.1580	9830	43
18	.50453	.86339	.58435	.7113	.1582	.9820	42
19	.50478	.86325	.58474	.7101	.1584	.9811	41
20	.50503	.86310	.58513	1.7090	1.1586	1.9801	40
21	.50528	.86295	.58552	.7079	.1588	9791	39
22	.50553	.86281	.58591	.7067	.1590	.9781	38
23	.50578	.86266	.58630	.7056	.1592	.9771	37
24	.50603	.86251	.58670	.7044	.1594	.9761	36
25	.50628	.86237	.58709	1.7033	1.1596	1.9752	35
26	.50653	.86222	58748	.7022	.1598	.9742	34
27	.50679	.86207	.58787	.7010	.1600	.9732	33
28	.50704	.86192	.58826	.6999	.1602	.9722	32
29	.50729	.86178	.58865	.6988	.1604	.9713	31
30	.50754	86163	.58904	1.6977	1.1606	1.9703	30
31	.50779	.86148	.58944	.6965	.1608	.9693	29
32	.50804	.86133	.58983	.6954	.1610	.9683	28
33	.50829	.86118	.59022	.6943	.1612	.9674	27
34	.50854	.86104	.59061	.6931	.1614	.9664	26
35	.50879	.86089	.59100	1.6920	1.1616	1.9654	25
36	.50904	.86074	.59140	.6909	.1618	.9645	24
37	.50929	.86059	.59179	.6898	.1620	.9635	23
38	.50954	.86044	.59218	.6887	.1622	.9625	22
39	.50979	.86030	.59258	.6875	.1624	.9616	21
40	.51004	.86015	.59297	1.6864	1.1626	1.9606	20
41	.51029	.86000	.59336	.6853	.1628	.9596	19
42	.51054	.85985	.59376	.6842	.1630	9587	18
43	.51079	.85970	59415	.6831	.1632	9577	17
44	.51104	.85955	.59454	.6820	.1634	.9568	16
45	.51129	.85941	.59494	1.6808	1.1636	1.9558	15
46	.51154	.85926	.59533	.6797	.1638	.9549	14
47	.51179	.85911	.59572	.6786	.1640	.9539	13
48	.51204	.85896	.59612	.6775	.1642	.9530	12
49	.51229	.85881	.59651	.6764	.1644	.9520	11
50	.51254	.85866	.59691	1.6753	1.1646	1.9510	10
51	.51279	.85851	.59730	.6742	.1648	.9501	9
52	.51304	.85836	.59770	.6731	.1650	.9491	8
53	.51329	.85821	.59809	.6720	.1652	9482	7
54	.51354	.85806	.59849	.6709	.1654	.9473	6
55	.51379	.85791	.59888	1.6698	1.1656	1.9463	5
56	.51404	.85777	.59928	.6687	.1658	.9454	4
57	.51429	.85762	.59967	.6676	.1660	.9444	3
58	.51454	.85747	.60007	.6665	.1662	.9435	2
59	.51479	.85732	.60046	.6654	.1664	.9425	1
60	51504	.85717	.60086	1.6643	1.1666	1.9416	0
M	Cosine	Sine	Cotan.	Tan.	Cosec.	Secant	M

59°

31°

M	Sine	Cosine	Tan.	Cotan.	Secant	Cosec.	M
0	.51504	.85717	.60086	1.6643	1.1666	1.9416	60
1	.51529	.85702	.60126	.6632	.1668	.9407	59
2	.51554	[illegible]	.60165	.6621	.1670	.9397	58
3	.51578	.85672	.60205	.6610	.1672	.9388	57
4	.51603	.85657	.60244	.6599	.1674	.9378	56
5	.51628	.85642	.60284	1.6588	1.1676	1.9369	55
6	.51653	.85627	.60324	.6577	.1678	.9360	54
7	.51678	.85612	.60363	.6566	.1681	.9350	53
8	.51703	.85597	.60403	.6555	.1683	.9341	52
9	.51728	.85582	.60443	.6544	.1685	.9332	51
10	.51753	.85566	.60483	1.6534	1.1687	1.9322	50
11	.51778	.85551	.60522	.6523	.1689	.9313	49
12	.51803	.85536	.60562	.6512	.1691	.9304	48
13	.51827	.85521	.60602	.6501	.1693	.9295	47
14	.51852	.85506	.60642	.6490	.1695	.9285	46
15	.51877	.85491	.60681	1.6479	1.1697	1.9276	45
16	.51902	.85476	.60721	.6469	.1699	.9267	44
17	.51927	.85461	.60761	.6458	.1701	.9258	43
18	.51952	.85446	.60801	.6447	.1703	.9248	42
19	.51977	.85431	.60841	.6436	.1705	.9239	41
20	.52002	.85416	.60881	1.6425	1.1707	1.9230	40
21	.52026	.85400	.60920	.6415	.1709	.9221	39
22	.52051	.85385	.60960	.6404	.1712	.9212	38
23	.52076	.85370	.61000	.6393	.1714	.9203	37
24	.52101	.85355	.61040	.6383	.1716	.9193	36
25	.52126	.85340	.61080	1.6372	1.1718	1.9184	35
26	.52151	.85325	.61120	.6361	.1720	.9175	34
27	.52175	.85309	.61160	.6350	.1722	.9166	33
28	.52200	.85294	.61200	.6340	.1724	.9157	32
29	.52225	.85279	.61240	.6329	.1726	.9148	31
30	.52250	.85264	.61280	1.6318	1.1728	1.9139	30
31	.52275	.85249	.61320	.6308	.1730	.9130	29
32	.52299	.85234	.61360	.6297	.1732	.9121	28
33	.52324	.85218	.61400	.6286	.1734	.9112	27
34	.52349	.85203	.61440	.6276	.1737	.9102	26
35	.52374	.85188	.61480	1.6265	1.1739	1.9093	25
36	.52398	.85173	.61520	.6255	.1741	.9084	24
37	.52423	.85157	.61560	.6244	.1743	.9075	23
38	.52448	.85142	.61601	.6233	.1745	.9066	22
39	.52473	.85127	.61641	.6223	.1747	.9057	21
40	.52498	.85112	.61681	1.6212	1.1749	1.9048	20
41	.52522	.85096	.61721	.6202	.1751	.9039	19
42	.52547	.85081	.61761	.6191	.1753	.9030	18
43	.52572	.85066	.61801	.6181	.1756	.9021	17
44	.52597	.85050	.61842	.6170	.1758	.9013	16
45	.52621	.85035	.61882	1.6160	1.1760	1.9004	15
46	.52646	.85020	.61922	.6149	.1762	.8995	14
47	.52671	.85004	.61962	.6139	.1764	.8986	13
48	.52695	.84989	.62003	.6128	.1766	.8977	12
49	.52720	.84974	.62043	.6118	.1768	.8968	11
50	.52745	.84959	.62083	1.6107	1.1770	1.8959	10
51	.52770	.84943	.62123	.6097	.1772	.8950	9
52	.52794	.84928	.62164	.6086	.1775	.8941	8
53	.52819	.84912	.62204	.6076	.1777	.8932	7
54	.52844	.84897	.62244	.6066	.1779	.8924	6
55	.52868	.84882	.62285	1.6055	1.1781	1.8915	5
56	.52893	.84866	.62325	.6045	.1783	.8906	4
57	.52918	.84851	.62366	.6034	.1785	.8897	3
58	.52942	.84836	.62406	.6024	.1787	.8888	2
59	.52967	.84820	.62446	.6014	.1790	.8879	1
60	.52992	.84805	.62487	1.6003	1.1792	1.8871	0
M	Cosine	Sine	Cotan.	Tan.	Cosec.	Secant	M

58°

32°

M	Sine	Cosine	Tan.	Cotan.	Secant	Cosec.	M
0	.52992	.84805	.62487	1.6003	1.1792	1.8871	60
1	.53016	.84789	.62527	.5993	.1794	.8862	59
2	.53041	.84774	.62568	.5983	.1796	.8853	58
3	.53066	.84758	.62608	.5972	.1798	.8844	57
4	.53090	.84743	.62649	.5962	.1800	.8836	56
5	.53115	.84728	.62689	1.5952	1.1802	1.8827	55
6	.53140	.84712	.62730	.5941	.1805	.8818	54
7	.53164	.84697	.62770	.5931	.1807	.8809	53
8	.53189	.84681	.62811	.5921	.1809	.8801	52
9	.53214	.84666	.62851	.5910	.1811	.8792	51
10	.53238	.84650	.62892	1.5900	1.1813	1.8783	50
11	.53263	.84635	.62933	.5890	.1815	.8775	49
12	.53288	.84619	.62973	.5880	.1818	.8766	48
13	.53312	.84604	.63014	.5869	.1820	.8757	47
14	.53337	.84588	.63055	.5859	.1822	.8749	46
15	.53361	.84573	.63095	1.5849	1.1824	1.8740	45
16	.53386	.84557	.63136	.5839	.1826	.8731	44
17	.53411	.84542	.63177	.5829	.1828	.8723	43
18	.53435	.84526	.63217	.5818	.1831	.8714	42
19	.53460	.84511	.63258	.5808	.1833	.8706	41
20	.53484	.84495	.63299	1.5798	1.1835	1.8697	40
21	.53509	.84479	.63339	.5788	.1837	.8688	39
22	.53533	.84464	.63380	.5778	.1839	.8680	38
23	.53558	.84448	.63421	.5768	.1841	.8671	37
24	.53583	.84433	.63462	.5757	.1844	.8663	36
25	.53607	.84417	.63503	1.5747	1.1846	1.8654	35
26	.53632	.84402	.63543	.5737	.1848	.8646	34
27	.53656	.84386	.63584	.5727	.1850	.8637	33
28	.53681	.84370	.63625	.5717	.1852	.8629	32
29	.53705	.84355	.63666	.5707	.1855	.8620	31
30	.53730	.84339	.63707	1.5697	1.1857	1.8611	30
31	.53754	.84323	.63748	.5687	.1859	.8603	29
32	.53779	.84308	.63789	.5677	.1861	.8595	28
33	.53803	.84292	.63830	.5667	1863	.8586	27
34	.53828	.84276	.638/1	.5657	.1866	.8578	26
35	.53852	.84261	.63912	1.5646	1.1868	1.8569	25
36	.53877	.84245	.63953	.5636	.1870	.8561	24
37	.53901	.84229	.63994	.5626	.1872	.8552	23
38	.53926	.84214	.64035	.5616	.1874	.8544	22
39	.53950	.84198	.64076	.5606	.1877	.8535	21
40	.53975	.84182	.64117	1.5596	1.1879	1.8527	20
41	.53999	.84167	.64158	.5586	.1881	.8519	19
42	.54024	.84151	.64199	.5577	.1883	.8510	18
43	.54048	.84135	.64240	.5567	.1886	.8502	17
44	.54073	.84120	.64281	.5557	.1888	.8493	16
45	.54097	.84104	.64322	1.5547	1.1890	1.8485	15
46	.54122	.84088	.64363	.5537	.1892	.8477	14
47	.54146	.84072	.64404	.5527	.1894	.8468	13
48	.54171	.84057	.64446	.5517	.1897	.8460	12
49	.54195	.84041	.64487	.5507	.1899	.8452	11
50	.54220	.84025	.64528	1.5497	1.1901	1.8443	10
51	.54244	.84009	.64569	.5487	.1903	.8435	9
52	.54268	.83993	.64610	.5477	.1906	.8427	8
53	.54293	.83978	.64652	.5467	.1908	.8418	7
54	.54317	.83962	.64693	.5458	.1910	.8410	6
55	.54342	.83946	.64734	1.5448	1.1912	1.8402	5
56	.54366	.83930	.64775	.5438	.1915	.8394	4
57	.54391	.83914	.64817	.5428	.1917	.8385	3
58	.54415	.83899	.64858	.5418	.1919	.8377	2
59	.54439	.83883	.64899	.5408	.1921	.8369	1
60	.54464	.83867	.64941	1.5399	1.1924	1.8361	0
M	Cosine	Sine	Cotan.	Tan.	Cosec.	Secant	M

57°

33°

M	Sine	Cosine	Tan.	Cotan.	Secant	Cosec.	M
0	.54464	.83867	.64941	1.5396	1.1924	1.8361	60
1	.54488	.83851	.64982	.5389	.1926	.8352	59
2	.54513	.83835	.65023	.5379	.1928	.8344	58
3	.54537	.83819	.65065	.5369	.1930	.8336	57
4	.54561	.83804	.65106	.5359	.1933	.8328	56
5	.54586	.83788	.65148	1.5350	1.1935	1.8320	55
6	.54610	.83772	.65189	.5340	.1937	.8311	54
7	.54634	.83756	.65231	.5330	.1939	.8303	53
8	.54659	.83740	.65272	.5320	.1942	.8295	52
9	.54683	.83724	.65314	.5311	.1944	.8287	51
10	.54708	.83708	.65355	1.5301	1.1946	1.8279	50
11	.54732	.83692	.65397	.5291	.1948	.8271	49
12	.54756	.83676	.65438	.5282	.1951	.8263	48
13	.54781	.83660	.65480	.5272	.1953	.8255	47
14	.54805	.83644	.65521	.5262	.1955	.8246	46
15	.54829	.83629	.65563	1.5252	1.1958	1.8238	45
16	.54854	.83613	.65604	.5243	.1960	.8230	44
17	.54878	.83597	.65646	.5233	.1962	.8222	43
18	.54902	.83581	.65688	.5223	.1964	.8214	42
19	.54926	.83565	.65729	.5214	.1967	.8206	41
20	.54951	.83549	.65771	1.5204	1.1969	1.8198	40
21	.54975	.83533	.65813	.5195	.1971	.8190	39
22	.54999	.83517	.65854	.5185	.1974	.8182	38
23	.55024	.83501	.65896	.5175	.1976	.8174	37
24	.55048	.83485	.65938	.5166	.1978	.8166	36
25	.55072	.83469	.65980	1.5156	1.1980	1.8158	35
26	.55097	.83453	.66021	.5147	.1983	.8150	34
27	.55121	.83437	.66063	.5137	.1985	.8142	33
28	.55145	.83421	.66105	.5127	.1987	.8134	32
29	.55169	.83405	.66147	.5118	.1990	.8126	31
30	.55194	.83388	.66188	1.5108	1.1992	1.8118	30
31	.55218	.83372	.66230	.5099	.1994	.8110	29
32	.55242	.83356	.66272	.5089	.1997	.8102	28
33	.55266	.83340	.66314	.5080	.1999	.8094	27
34	.55291	.83324	.66356	.5070	.2001	.8086	26
35	.55315	.83308	.66398	1.5061	1.2004	1.8078	25
36	.55339	.83292	.66440	.5051	.2006	.8070	24
37	.55363	.83276	.66482	.5042	.2008	.8062	23
38	.55388	.83260	.66524	.5032	.2010	.8054	22
39	.55412	.83244	.66566	.5023	.2013	.8047	21
40	.55436	.83228	.66608	1.5013	1.2015	1.8039	20
41	.55460	.83211	.66650	.5004	.2017	.8031	19
42	.55484	.83195	.66692	.4994	.2020	.8023	18
43	.55509	.83179	.66734	.4985	.2022	.8015	17
44	.55533	.83163	.66776	.4975	.2024	.8007	16
45	.55557	.83147	.66818	1.4966	1.2027	1.7999	15
46	.55581	.83131	.66860	.4957	.2029	.7992	14
47	.55605	.83115	.66902	.4947	.2031	.7984	13
48	.55629	.83098	.66944	.4938	.2034	.7976	12
49	.55654	.83082	.66986	.4928	.2036	.7968	11
50	.55678	.83066	.67028	1.4919	1.2039	1.7960	10
51	.55702	.83050	.67071	.4910	.2041	.7953	9
52	.55726	.83034	.67113	.4900	.2043	.7945	8
53	.55750	.83017	.67155	.4891	.2046	.7937	7
54	.55774	.83001	.67197	.4881	.2048	.7929	6
55	.55799	.82985	.67239	1.4872	1.2050	1.7921	5
56	.55823	.82969	.67282	.4863	.2053	.7914	4
57	.55847	.82952	.67324	.4853	.2055	.7906	3
58	.55871	.82936	.67366	.4844	.2057	.7898	2
59	.55895	.82920	.67408	.4835	.2060	.7891	1
60	.55919	.82904	.67451	1.4826	1.2062	1.7883	0
M	Cosine	Sine	Cotan.	Tan.	Cosec.	Secant	M

56°

34°

M	Sine	Cosine	Tan.	Cotan.	Secant	Cosec.	M
0	.55919	.82904	.67451	1.4826	1.2062	1.7883	60
1	.55943	.82887	.67493	.4816	.2064	.7875	59
2	.55967	.82871	.67535	.4807	.2067	.7867	58
3	.55992	.82855	.67578	.4798	.2069	.7860	57
4	.56016	.82839	.67620	.4788	.2072	.7852	56
5	.56040	.82822	.67663	1.4779	1.2074	1.7844	55
6	.56064	.82806	.67705	.4770	.2076	.7837	54
7	.56088	.82790	.67747	.4761	.2079	.7829	53
8	.56112	.82773	.67790	.4751	.2081	.7821	52
9	.56136	.82757	.67832	.4742	.2083	.7814	51
10	.56160	.82741	.67875	1.4733	1.2086	1.7806	50
11	.56184	.82724	.67917	.4724	.2088	.7798	49
12	.56208	.82708	.67960	.4714	.2091	.7791	48
13	.56232	.82692	.68002	.4705	.2093	.7783	47
14	.56256	.82675	.68045	.4696	.2095	.7776	46
15	.56280	.82659	.68087	1.4687	1.2098	1.7768	45
16	.56304	.82643	.68130	.4678	.2100	.7760	44
17	.56328	.82626	.68173	.4669	.2103	.7753	43
18	.56353	.82610	.68215	.4659	.2105	.7745	42
19	.56377	.82593	.68258	.4650	.2107	.7738	41
20	.56401	.82577	.68301	1.4641	1.2110	1.7730	40
21	.56425	.82561	.68343	.4632	.2112	.7723	39
22	.56449	.82544	.68386	.4623	.2115	.7715	38
23	.56473	.82528	.68429	.4614	.2117	.7708	37
24	.56497	.82511	.68471	.4605	.2119	.7700	36
25	.56521	.82495	.68514	1.4595	1.2122	1.7693	35
26	.56545	.82478	.68557	.4586	.2124	.7685	34
27	.56569	.82462	.68600	.4577	.2127	.7678	33
28	.56593	.82445	.68642	.4568	.2129	.7670	32
29	.56617	.82429	.68685	.4559	.2132	.7663	31
30	.56641	.82413	.68728	1.4550	1.2134	1.7655	30
31	.56664	.82396	.68771	.4541	.2136	.7648	29
32	.56688	.82380	.68814	.4532	.2139	.7640	28
33	.56712	.82363	.68857	.4523	.2141	.7633	27
34	.56736	.82347	.68899	.4514	.2144	.7625	26
35	.56760	.82330	.68942	1.4505	1.2146	1.7618	25
36	.56784	.82314	.68985	.4496	.2149	.7610	24
37	.56808	.82297	.69028	.4487	.2151	.7603	23
38	.56832	.82280	.69071	.4478	.2153	.7596	22
39	.56856	.82264	.69114	.4469	.2156	.7588	21
40	.56880	.82247	.69157	1.4460	1.2158	1.7581	20
41	.56904	.82231	.69200	.4451	.2161	.7573	19
42	.56928	.82214	.69243	.4442	.2163	.7566	18
43	.56952	.82198	.69286	.4433	.2166	.7559	17
44	.56976	.82181	.69329	.4424	.2168	.7551	16
45	.57000	.82165	.69372	1.4415	1.2171	1.7544	15
46	.57023	.82148	.69415	.4406	.2173	.7537	14
47	.57047	.82131	.69459	.4397	.2175	.7529	13
48	.57071	.82115	.69502	.4388	.2178	.7522	12
49	.57095	.82098	.69545	.4379	.2180	.7514	11
50	.57119	.82082	.69588	1.4370	1.2183	1.7507	10
51	.57143	.82065	.69631	.4361	.2185	.7500	9
52	.57167	.82048	.69674	.4352	.2188	.7493	8
53	.57191	.82032	.69718	.4343	.2190	.7485	7
54	.57214	.82015	.69761	.4335	.2193	.7478	6
55	.57238	.81998	.69804	1.4326	1.2195	1.7471	5
56	.57262	.81982	.69847	.4317	.2198	.7463	4
57	.57286	.81965	.69891	.4308	.2200	.7456	3
58	.57310	.81948	.69934	.4299	.2203	.7449	2
59	.57334	.81932	.69977	.4290	.2205	.7442	1
60	.57358	.81915	.70021	1.4281	1.2208	1.7434	0
M	Cosine	Sine	Cotan.	Tan.	Cosec.	Secant	M

55°

35°

M	Sine	Cosine	Tan.	Cotan.	Secant	Cosec.	M
0	.57358	.81915	.70021	1.4281	1.2208	1.7434	60
1	.57381	.81898	.70064	.4273	.2210	.7427	59
2	.57405	.81882	.70107	.4264	.2213	.7420	58
3	.57429	.81865	.70151	.4255	.2215	.7413	57
4	.57453	.81848	.70194	.4246	.2218	.7405	56
5	.57477	.81832	.70238	1.4237	1.2220	1.7398	55
6	.57500	.81815	.70281	.4228	.2223	.7391	54
7	.57524	.81798	.70325	.4220	.2225	.7384	53
8	.57548	.81781	.70368	.4211	.2228	.7377	52
9	.57572	.81765	.70412	.4202	.2230	.7369	51
10	.57596	.81748	.70455	1.4193	1.2233	1.7362	50
11	.57619	.81731	.70499	.4185	.2235	.7355	49
12	.57643	.81714	.70542	.4176	.2238	.7348	48
13	.57667	.81698	.70586	.4167	.2240	.7341	47
14	.57691	.81681	.70629	.4158	.2243	.7334	46
15	.57714	.81664	.70673	1.4150	1.2245	1.7327	45
16	.57738	.81647	.70717	.4141	.2248	.7319	44
17	.57762	.81630	.70760	.4132	.2250	.7312	43
18	.57786	.81614	.70804	.4123	.2253	.7305	42
19	.57809	.81597	.70848	.4115	.2255	.7298	41
20	.57833	.81580	.70891	1.4106	1.2258	1.7291	40
21	.57857	.81563	.70935	.4097	.2260	.7284	39
22	.57881	.81546	.70979	.4089	.2263	.7277	38
23	.57904	.81530	.71022	.4080	.2265	.7270	37
24	.57928	.81513	.71066	.4071	.2268	.7263	36
25	.57952	.81496	.71110	1.4063	1.2270	1.7256	35
26	.57975	.81479	.71154	.4054	.2273	.7249	34
27	.57999	.81462	.71198	.4045	.2276	.7242	33
28	.58023	.81445	.71241	1.4037	.2278	.7234	32
29	.58047	.81428	.71285	.4028	.2281	.7227	31
30	.58070	.81411	.71329	1.4019	1.2283	1.7220	30
31	.58094	.81395	.71373	.4011	.2286	.7213	29
32	.58118	.81378	.71417	.4002	.2288	.7206	28
33	.58141	.81361	.71461	.3994	.2291	.7199	27
34	.58165	.81344	.71505	.3985	.2293	.7192	26
35	.58189	.81327	.71549	1.3976	1.2296	1.7185	25
36	.58212	.81310	.71593	.3968	.2298	.7178	24
37	.58236	.81293	.71637	.3959	1.2301	.7171	23
38	.58259	.81276	.71681	.3951	.2304	.7164	22
39	.58283	.81259	.71725	.3942	.2306	.7157	21
40	.58307	.81242	.71769	1.3933	1.2309	1.7151	20
41	.58330	.81225	.71813	.3925	.2311	.7144	19
42	.58354	.81208	.71857	.3916	.2314	.7137	18
43	.58378	.81191	.71901	.3908	.2316	.7130	17
44	.58401	.81174	.71945	.3899	.2319	.7123	16
45	.58425	.81157	.71990	1.3891	1.2322	1.7116	15
46	.58448	.81140	.72034	.3882	.2324	.7109	14
47	.58472	.81123	.72078	.3874	.2327	.7102	13
48	.58496	.81106	.72122	.3865	.2329	.7095	12
49	.58519	.81089	.72166	.3857	.2332	.7088	11
50	.58543	.81072	.72211	1.3848	1.2335	1.7081	10
51	.58566	.81055	.72255	.3840	.2337	.7075	9
52	.58590	.81038	.72299	1.3831	.2340	.7068	8
53	.58614	.81021	.72344	.3823	.2342	.7061	7
54	.58637	.81004	.72388	.3814	.2345	.7054	6
55	.58661	.80987	.72432	1.3806	1.2348	1.7047	5
56	.58684	.80970	.72477	.3797	.2350	.7040	4
57	.58708	.80953	.72521	.3789	.2353	.7033	3
58	.58731	.80936	.72565	.3781	.2355	.7027	2
59	.58755	.80919	.72610	.3772	.2358	.7020	1
60	.58778	.80902	.72654	1.3764	1.2361	1.7013	0
M	Cosine	Sine	Cotan.	Tan.	Cosec.	Secant	M

54°

36°

M	Sine	Cosine	Tan.	Cotan.	Secant	Cosec.	M
0	.58778	.80902	.72654	1.3764	1.2361	1.7013	60
1	.58802	.80885	.72699	.3755	.2363	.7006	59
2	.58825	.80867	.72743	.3747	.2366	.6999	58
3	.58849	.80850	.72788	.3738	.2368	.6993	57
4	.58873	.80833	.72832	1.3730	.2371	.6986	56
5	.58896	.80816	.72877	1.3722	1.2374	1.6979	55
6	.58920	.80799	.72921	.3713	.2376	.6972	54
7	.58943	.80782	.72966	.3705	.2379	.6965	53
8	.58967	.80765	.73010	.3697	.2382	.6959	52
9	.58990	.80747	.73055	.3688	.2384	.6952	51
10	.59014	.80730	.73100	1.3680	1.2387	1.6945	50
11	.59037	.80713	.73144	.3672	.2389	.6938	49
12	.59060	.80696	.73189	.3663	.2392	.6932	48
13	.59084	.80679	.73234	.3655	.2395	.6925	47
14	.59107	.80662	.73278	.3647	.2397	.6918	46
15	.59131	.80644	.73323	1.3638	1.2400	1.6912	45
16	.59154	.80627	.73368	.3630	.2403	.6905	44
17	.59178	.80610	.73412	.3622	.2405	.6898	43
18	.59201	.80593	.73457	.3613	.2408	.6891	42
19	.59225	.80576	.73502	.3605	.2411	.6885	41
20	.59248	.80558	.73547	1.3597	1.2413	1.6878	40
21	.59272	.80541	.73592	.3588	.2416	.6871	39
22	.59295	.80524	.73637	.3580	.2419	.6865	38
23	.59318	.80507	.73681	.3572	.2421	.6858	37
24	.59342	.80489	.73726	.3564	.2424	.6851	36
25	.59365	.80472	.73771	1.3555	1.2427	1.6845	35
26	.59389	.80455	.73816	.3547	.2429	1.6838	34
27	.59412	.80437	.73861	.3539	.2432	.6831	33
28	.59435	.80420	.73906	.3531	.2435	.6825	32
29	.59459	.80403	.73951	.3522	.2437	.6818	31
30	.59482	.80386	.73996	1.3514	1.2440	1.6812	30
31	.59506	.80368	.74041	.3506	.2443	.6805	29
32	.59529	.80351	.74086	.3498	.2445	.6798	28
33	.59552	.80334	.74131	.3489	.2448	.6792	27
34	.59576	.80316	.74176	.3481	.2451	.6785	26
35	.59599	.80299	.74221	1.3473	1.2453	1.6779	25
36	.59622	.80282	.74266	.3465	.2456	.6772	24
37	.59646	.80264	.74312	.3457	.2459	.6766	23
38	.59669	.80247	.74357	.3449	.2461	.6759	22
39	.59692	.80230	.74402	.3440	.2464	.6752	21
40	.59716	.80212	.74447	1.3432	1.2467	1.6746	20
41	.59739	.80195	.74492	.3424	.2470	.6739	19
42	.59762	.80177	.74538	.3416	.2472	.6733	18
43	.59786	.80160	.74583	.3408	.2475	.6726	17
44	.59809	.80143	.74628	.3400	.2478	.6720	16
45	.59832	.80125	.74673	1.3392	1.2480	1.6713	15
46	.59856	.80108	.74719	1.3383	.2483	.6707	14
47	.59879	.80090	.74764	.3375	.2486	.6700	13
48	.59902	.80073	.74809	.3367	.2488	.6694	12
49	.59926	.80056	.74855	.3359	.2491	.6687	11
50	.59949	.80038	.74900	1.3351	1.2494	1.6681	10
51	.59972	.80021	.74946	.3343	.2497	.6674	9
52	.59995	.80003	.74991	.3335	.2499	.6668	8
53	.60019	.79986	.75037	.3327	.2502	.6661	7
54	.60042	.79968	.75082	.3319	.2505	.6655	6
55	.60065	.79951	.75128	1.3311	1.2508	1.6648	5
56	.60088	.79933	.75173	.3303	.2510	.6642	4
57	.60112	.79916	.75219	.3294	.2513	.6636	3
58	.60135	.79898	.75264	.3286	.2516	.6629	2
59	.60158	.79881	.75310	.3278	.2519	.6623	1
60	.60181	.79863	.75355	1.3270	1.2521	1.6616	0
M	Cosine	Sine	Cotan.	Tan.	Cosec.	Secant	M

53°

37°

M	Sine	Cosine	Tan.	Cotan.	Secant	Cosec.	M
0	.60181	.79863	.75355	1.3270	1.2521	1.6616	60
1	.60205	.79846	.75401	.3262	.2524	.6610	59
2	.60228	.79828	.75447	.3254	.2527	.6603	58
3	.60251	.79811	.75492	.3246	.2530	.6597	57
4	.60274	.79793	.75538	.3238	.2532	.6591	56
5	.60298	.79776	.75584	1.3230	1.2535	1.6584	55
6	.60320	.79758	.75629	.3222	.2538	.6578	54
7	.60344	.79741	.75675	.3214	.2541	.6572	53
8	.60367	.79723	.75721	.3206	.2543	.6565	52
9	.60390	.79706	.75767	.3198	.2546	.6559	51
10	.60413	.79688	.75812	1.3190	1.2549	1.6552	50
11	.60437	.79670	.75858	.3182	.2552	.6546	49
12	.60460	.79653	.75904	.3174	.2554	.6540	48
13	.60483	.79635	.75950	.3166	.2557	.6533	47
14	.60506	.79618	.75996	.3159	.2560	.6527	46
15	.60529	.79600	.76042	1.3151	1.2563	1.6521	45
16	.60552	.79582	.76088	.3143	.2565	.6514	44
17	.60576	.79565	.76134	.3135	.2568	.6508	43
18	.60599	.79547	.76179	.3127	.2571	.6502	42
19	.60622	.79530	.76225	.3119	.2574	.6496	41
20	.60645	.79512	.76271	1.3111	1.2577	1.6489	40
21	.60668	.79494	.76317	.3103	.2579	.6483	39
22	.60691	.79477	.76364	.3095	.2582	.6477	38
23	.60714	.79459	.76410	.3087	.2585	.6470	37
24	.60737	.79441	.76456	.3079	.2588	.6464	36
25	.60761	.79424	.76502	1.3071	1.2591	1.6458	35
26	.60784	.79406	.76548	.3064	.2593	.6452	34
27	.60807	.79388	.76594	.3056	.2596	.6445	33
28	.60830	.79371	.76640	.3048	.2599	.6439	32
29	.60853	.79353	.76686	.3040	.2602	.6433	31
30	.60876	.79335	.76733	1.3032	1.2605	1.6427	30
31	.60899	.79318	.76779	.3024	.2607	.6420	29
32	.60922	.79300	.76825	.3016	.2610	.6414	28
33	.60945	.79282	.76871	.3009	.2613	.6408	27
34	.60968	.79264	.76918	.3001	.2616	.6402	26
35	.60991	.79247	.76964	1.2993	1.2619	1.6396	25
36	.61014	.79229	.77010	.2985	.2622	.6389	24
37	.61037	.79211	.77057	.2977	.2624	.6383	23
38	.61061	.79193	.77103	.2970	.2627	.6377	22
39	.61084	.79176	.77149	.2962	.2630	.6371	21
40	.61107	.79158	.77196	1.2954	1.2633	1.6365	20
41	.61130	.79140	.77242	.2946	.2636	.6359	19
42	.61153	.79122	.77289	.2938	.2639	.6352	18
43	.61176	.79104	.77335	.2931	.2641	.6346	17
44	.61199	.79087	.77382	.2923	.2644	.6340	16
45	.61222	.79069	.77428	1.2915	1.2647	1.6334	15
46	.61245	.79051	.77475	.2907	.2650	.6328	14
47	.61268	.79033	.77521	.2900	.2653	.6322	13
48	.61290	.79015	.77568	.2892	.2656	.6316	12
49	.61314	.78998	.77614	.2884	.2659	.6309	11
50	.61337	.78980	.77661	1.2876	1.2661	1.6303	10
51	.61360	.78962	.77708	.2869	.2664	.6297	9
52	.61383	.78944	.77754	.2861	.2667	.6291	8
53	.61405	.78926	.77801	.2853	.2670	.6285	7
54	.61428	.78908	.77848	.2845	.2673	.6279	6
55	.61451	.78890	.77895	1.2838	1.2676	1.6273	5
56	.61474	.78873	.77941	.2830	.2679	.6267	4
57	.61497	.78855	.77988	.2822	.2681	.6261	3
58	.61520	.78837	.78035	.2815	.2684	.6255	2
59	.61543	.78819	.78082	.2807	.2687	.6249	1
60	.61566	.78801	.78128	1.2799	1.2690	1.6243	0
M	Cosine	Sine	Cotan.	Tan.	Cosec.	Secant	M

52°

38°

M	Sine	Cosine	Tan.	Cotan.	Secant	Cosec.	M
0	.61566	.78801	.78128	1.2799	1.2690	1.6243	60
1	.61589	.78783	.78175	.2792	.2693	.6237	59
2	.61612	.78765	.78222	.2784	.2696	.6231	58
3	.61635	.78747	.78269	.2776	.2699	.6224	57
4	.61658	.78729	.78316	.2769	.2702	.6218	56
5	.61681	.78711	.78363	1.2761	1.2705	1.6212	55
6	.61703	.78693	.78410	.2753	.2707	.6206	54
7	.61726	.78675	.78457	.2746	.2710	.6200	53
8	.61749	.78657	.78504	.2738	.2713	.6194	52
9	.61772	.78640	.78551	.2730	.2716	.6188	51
10	.61795	.78622	.78598	1.2723	1.2719	1.6182	50
11	.61818	.78604	.78645	.2715	.2722	.6176	49
12	.61841	.78586	.78692	.2708	.2725	.6170	48
13	.61864	.78568	.78739	.2700	.2728	.6164	47
14	.61886	.78550	.78786	.2692	.2731	.6159	46
15	.61909	.78532	.78834	1.2685	1.2734	1.6153	45
16	.61932	.78514	.78881	.2677	.2737	.6147	44
17	.61955	.78496	.78928	.2670	.2739	.6141	43
18	.61978	.78478	.78975	.2662	.2742	.6135	42
19	.62001	.78460	.79022	.2655	.2745	.6129	41
20	.62023	.78441	.79070	1.2647	1.2748	1.6123	40
21	.62046	.78423	.79117	.2639	.2751	.6117	39
22	.62069	.78405	.79164	.2632	.2754	.6111	38
23	.62092	.78387	.79212	.2624	.2757	.6105	37
24	.62115	.78369	.79259	.2617	.2760	.6099	36
25	.62137	.78351	.79306	1.2609	1.2763	1.6093	35
26	.62160	.78333	.79354	.2602	.2766	.6087	34
27	.62183	.78315	.79401	.2594	.2769	.6081	33
28	.62206	.78297	.79449	.2587	.2772	.6077	32
29	.62229	.78279	.79496	.2579	.2775	.6070	31
30	.62251	.78261	.79543	1.2572	1.2778	1.6064	30
31	.62274	.78243	.79591	.2564	.2781	.6058	29
32	.62297	.78224	.79639	.2557	.2784	.6052	28
33	.62320	.78206	.79686	.2549	.2787	.6046	27
34	.62342	.78188	.79734	.2542	.2790	.6040	26
35	.62365	.78170	.79781	1.2534	1.2793	1.6034	25
36	.62388	.78152	.79829	.2527	.2795	.6029	24
37	.62411	.78134	.79876	.2519	.2798	.6023	23
38	.62433	.78116	.79924	.2512	.2801	.6017	22
39	.62456	.78097	.79972	.2504	.2804	.6011	21
40	.62479	.78079	.80020	1.2497	1.2807	1.6005	20
41	.62501	.78061	.80067	.2489	.2810	.6000	19
42	.62524	.78043	.80115	.2482	.2813	.5994	18
43	.62547	.78025	.80163	.2475	.2816	.5988	17
44	.62570	.78007	.80211	.2467	.2819	.5982	16
45	.62592	.77988	.80258	1.2460	1.2822	1.5976	15
46	.62615	.77970	.80306	.2452	.2825	.5971	14
47	.62638	.77952	.80354	.2445	.2828	.5965	13
48	.62660	.77934	.80402	.2437	.2831	.5959	12
49	.62683	.77915	.80450	.2430	.2834	.5953	11
50	.62706	.77897	.80498	1.2423	1.2837	1.5947	10
51	.62728	.77879	.80546	.2415	.2840	.5942	9
52	.62751	.77861	.80594	.2408	.2843	.5936	8
53	.62774	.77842	.80642	.2400	.2846	.5930	7
54	.62796	.77824	.80690	.2393	.2849	.5924	6
55	.62819	.77806	.80738	1.2386	1.2852	1.5919	5
56	.62841	.77788	.80786	.2378	.2855	.5913	4
57	.62864	.77769	.80834	.2371	.2858	.5907	3
58	.62887	.77751	.80882	.2364	.2861	.5901	2
59	.62909	.77733	.80930	.2356	.2864	.5896	1
60	.62932	.77715	.80978	1.2349	1.2867	1.5890	0
M	Cosine	Sine	Cotan.	Tan.	Cosec.	Secant	M

51°

39°

M	Sine	Cosine	Tan.	Cotan.	Secant	Cosec.	M
0	.62932	.77715	.80978	1.2349	1.2867	1.5890	60
1	.62955	.77696	.81026	.2342	.2871	.5884	59
2	.62977	.77678	.81075	.2334	.2874	.5879	58
3	.63000	.77660	.81123	.2327	.2877	.5873	57
4	.63022	.77641	.81171	.2320	.2880	.5867	56
5	.63045	.77623	.81219	1.2312	1.2883	1.5862	55
6	.63067	.77605	.81268	.2305	.2886	.5856	54
7	.63090	.77586	.81316	.2297	.2889	.5850	53
8	.63113	.77568	.81364	.2290	.2892	.5845	52
9	.63135	.77549	.81413	.2283	.2895	.5839	51
10	.63158	.77531	.81461	1.2276	1.2898	1.5833	50
11	.63180	.77513	.81509	.2268	.2901	.5828	49
12	.63203	.77494	.81558	.2261	.2904	.5822	48
13	.63225	.77476	.81606	.2254	.2907	.5816	47
14	.63248	.77458	.81655	.2247	.2910	.5811	46
15	.63270	.77439	.81703	1.2239	1.2913	1.5805	45
16	.63293	.77421	.81752	.2232	.2916	.5799	44
17	.63315	.77402	.81900	.2225	.2919	.5794	43
18	.63338	.77384	.81849	.2218	.2922	.5788	42
19	.63360	.77365	.81898	.2210	.2926	.5783	41
20	.63383	.77347	.81946	1.2203	1.2929	1.5777	40
21	.63405	.77329	.81995	.2196	.2932	.5771	39
22	.63428	.77310	.82043	.2189	.2935	.5766	38
23	.63450	.77292	.82092	.2181	.2938	.5760	37
24	.63473	.77273	.82141	.2174	.2941	.5755	36
25	.63495	.77255	.82190	1.2167	1.2944	1.5749	35
26	.63518	.77236	.82238	.2160	.2947	.5743	34
27	.63540	.77218	.82287	.2152	.2950	.5738	33
28	.63563	.77199	.82336	.2145	.2953	.5732	32
29	.63585	.77181	.82385	.2138	.2956	.5727	31
30	.63608	.77162	.82434	1.2131	1.2960	1.5721	30
31	.63630	.77144	.82482	.2124	.2963	.5716	29
32	.63653	.77125	.82531	.2117	.2966	.5710	28
33	.63675	.77107	.82580	.2109	.2969	.5705	27
34	.63697	.77088	.82629	.2102	.2972	.5699	26
35	.63720	.77070	.82678	1.2095	1.2975	1.5694	25
36	.63742	.77051	.82727	.2088	.2978	.5688	24
37	.63765	.77033	.82776	.2081	.2981	.5683	23
38	.63787	.77014	.82825	.2074	.2985	.5677	22
39	.63810	.76996	.82874	.2066	.2988	.5672	21
40	.63832	.76977	.82923	1.2059	1.2991	1.5666	20
41	.63854	.76958	.82972	.2052	.2994	.5661	19
42	.63877	.76940	.83022	.2045	.2997	.5655	18
43	.63899	.76921	.83071	.2038	.3000	.5650	17
44	.63921	.76903	.83120	.2031	.3003	.5644	16
45	.63944	.76884	.83169	1.2024	1.3006	1.5639	15
46	.63966	.76865	.83218	.2016	.3010	.5633	14
47	.63989	.76847	.83267	.2009	.3013	.5628	13
48	.64011	.76828	.83317	.2002	.3016	.5622	12
49	.64033	.76810	.83366	.1995	.3019	.5617	11
50	.64056	.76791	.83415	1.1988	1.3022	1.5611	10
51	.64078	.76772	.83465	.1981	.3025	.5606	9
52	.64100	.76754	.83514	.1974	.3029	.5600	8
53	.64123	.76735	.83563	.1967	.3032	.5595	7
54	.64145	.76716	.83613	.1960	.3035	.5590	6
55	.64167	.76698	.83662	1.1953	1.3038	1.5584	5
56	.64189	.76679	.83712	.1946	.3041	.5579	4
57	.64212	.76660	.83761	.1939	.3044	.5573	3
58	.64234	.76642	.83811	.1932	.3048	.5568	2
59	.64256	.76623	.83860	.1924	.3051	.5563	1
60	.64279	.76604	.83910	1.1917	1.3054	1.5557	0
M	Cosine	Sine	Cotan.	Tan.	Cosec.	Secant	M

50°

40°

M	Sine	Cosine	Tan.	Cotan.	Secant	Cosec.	M
0	.64279	.76604	.83910	1.1917	1.3054	1.5557	60
1	.64301	.76586	.83959	.1910	.3057	.5552	59
2	.64323	.76567	.84009	.1903	.3060	.5546	58
3	.64345	.76548	.84059	.1896	.3064	.5541	57
4	.64368	.76530	.84108	.1889	.3067	.5536	56
5	.64390	.76511	.84158	1.1882	1.3070	1.5530	55
6	.64412	.76492	.84208	.1875	.3073	.5525	54
7	.64435	.76473	.84257	.1868	.3076	.5520	53
8	.64457	.76455	.84307	.1861	.3080	.5514	52
9	.64479	.76436	.84357	.1854	.3083	.5509	51
10	.64501	.76417	.84407	1.1847	1.3086	1.5503	50
11	.64523	.76398	.84457	.1840	.3089	.5498	49
12	.64546	.76380	.84506	.1833	.3092	.5493	48
13	.64568	.76361	.84556	.1826	.3096	.5487	47
14	.64590	.76342	.84606	.1819	.3099	.5482	46
15	.64612	.76323	.84656	1.1812	1.3102	1.5477	45
16	.64635	.76304	.84706	.1806	.3105	.5471	44
17	.64657	.76286	.84756	.1798	.3109	.5466	43
18	.64679	.76267	.84806	.1791	.3112	.5461	42
19	.64701	.76248	.84856	.1785	.3115	.5456	41
20	.64723	.76229	.84906	1.1778	1.3118	1.5450	40
21	.64745	.76210	.84956	.1771	.3121	.5445	39
22	.64768	.76191	.85006	.1764	.3125	.5440	38
23	.64790	.76173	.85056	.1757	.3128	.5434	37
24	.64812	.76154	.85107	.1750	.3131	.5429	36
25	.64834	.76135	.85157	1.1743	1.3134	1.5424	35
26	.64856	.76116	.85207	.1736	.3138	.5419	34
27	.64878	.76097	.85257	.1729	.3141	.5413	33
28	.64900	.76078	.85307	.1722	.3144	.5408	32
29	.64923	.76059	.85358	.1715	.3148	.5403	31
30	.64945	.76041	.85408	1.1708	1.3151	1.5398	30
31	.64967	.76022	.85458	.1702	.3154	.5392	29
32	.64989	.76003	.85509	.1695	.3157	.5387	28
33	.65011	.75984	.85559	.1688	.3161	.5382	27
34	.65033	.75965	.85609	.1681	.3164	.5377	26
35	.65055	.75946	.85660	1.1674	1.3167	1.5371	25
36	.65077	.75927	.85710	.1667	.3170	.5366	24
37	.65100	.75908	.85761	.1660	.3174	.5361	23
38	.65121	.75889	.85811	.1653	.3177	.5356	22
39	.65144	.75870	.85862	.1647	.3180	.5351	21
40	.65166	.75851	.85912	1.1640	1.3184	1.5345	20
41	.65188	.75832	.85963	.1633	.3187	.5340	19
42	.65210	.75813	.86013	.1626	.3190	.5335	18
43	.65232	.75794	.86064	.1619	.3193	.5330	17
44	.65254	.75775	.86115	.1612	.3197	.5325	16
45	.65276	.75756	.86165	1.1605	1.3200	1.5319	15
46	.65298	.75737	.86216	.1599	.3203	.5314	14
47	.65320	.75718	.86267	.1592	.3207	.5309	13
48	.65342	.75700	.86318	.1585	.3210	.5304	12
49	.65364	.75680	.86368	.1578	.3213	.5290	11
50	.65386	.75661	.86419	1.1571	1.3217	1.5294	10
51	.65408	.75642	.86470	.1565	.3220	.5289	9
52	.65430	.75623	.86521	.1558	.3223	.5283	8
53	.65452	.75604	.86572	.1551	.3227	.5278	7
54	.65474	.75585	.86623	.1544	.3230	.5273	6
55	.65496	.75566	.86674	1.1537	1.3233	1.5268	5
56	.65518	.75547	.86725	.1531	.3237	.5263	4
57	.65540	.75528	.86775	.1524	.3240	.5258	3
58	.65562	.75509	.86826	.1517	.3243	.5253	2
59	.65584	.75490	.86878	.1510	.3247	.5248	1
60	.65606	.75471	.86929	1.1504	1.3250	1.5242	0
M	Cosine	Sine	Cotan.	Tan.	Cosec.	Secant	M

49°

41°

M	Sine	Cosine	Tan.	Cotan.	Secant	Cosec.	M
0	.65606	.75471	.86929	1.1504	1.3250	1.5242	60
1	.65628	.75452	.86980	.1497	.3253	.5237	59
2	.65650	.75433	.87031	.1490	.3257	.5232	58
3	.65672	.75414	.87082	.1483	.3260	.5227	57
4	.65694	.75394	.87133	.1477	.3263	.5222	56
5	.65716	.75375	.87184	1.1470	1.3267	1.5217	55
6	.65737	.75356	.87235	.1463	.3270	.5212	54
7	.65759	.75337	.87287	.1456	.3274	.5207	53
8	.65781	.75318	.87338	.1450	.3277	.5202	52
9	.65803	.75299	.87389	.1443	.3280	.5197	51
10	.65825	.75280	.87441	1.1436	1.3284	1.5192	50
11	.65847	.75261	.87492	.1430	.3287	.5187	49
12	.65869	.75241	.87543	.1423	.3290	.5182	48
13	.65891	.75222	.87595	.1416	.3294	.5177	47
14	.65913	.75203	.87646	.1409	.3297	.5171	46
15	.65934	.75184	.87698	1.1403	1.3301	1.5166	45
16	.65956	.75165	.87749	.1396	.3304	.5161	44
17	.65978	.75146	.87801	.1389	.3307	.5156	43
18	.66000	.75126	.87852	.1383	.3311	.5151	42
19	.66022	.75107	.87904	.1376	.3314	.5146	41
20	.66044	.75088	.87955	1.1369	1.3318	1.5141	40
21	.66066	.75069	.88007	.1363	.3321	.5136	39
22	.66087	.75049	.88058	.1356	.3324	.5131	38
23	.66109	.75030	.88110	.1349	.3328	.5126	37
24	.66131	.75011	.88162	.1343	.3331	.5121	36
25	.66153	.74992	.88213	1.1336	1.3335	1.5116	35
26	.66175	.74973	.88265	.1329	.3338	.5111	34
27	.66197	.74953	.88317	.1323	.3342	.5106	33
28	.66218	.74934	.88369	.1316	.3345	.5101	32
29	.66240	.74915	.88421	.1309	.3348	.5096	31
30	.66262	.74895	.88472	1.1303	1.3352	1.5092	30
31	.66284	.74876	.88524	.1296	.3355	.5087	29
32	.66305	.74857	.88576	.1290	.3359	.5082	28
33	.66327	.74838	.88628	.1283	.3362	.5077	27
34	.66349	.74818	.88680	.1276	.3366	.5072	26
35	.66371	.74799	.88732	1.1270	1.3369	1.5067	25
36	.66393	.74780	.88784	.1263	.3372	.5062	24
37	.66414	.74760	.88836	.1257	.3376	.5057	23
38	.66436	.74741	.88888	.1250	.3379	.5052	22
39	.66458	.74722	.88940	.1243	.3383	.5047	21
40	.66479	.74702	.88992	1.1237	1.3386	1.5042	20
41	.66501	.74683	.89044	.1230	.3390	.5037	19
42	.66523	.74664	.89097	.1224	.3393	.5032	18
43	.66545	.74644	.89149	.1217	.3397	.5027	17
44	.66566	.74625	.89201	.1211	.3400	.5022	16
45	.66588	.74606	.89253	1.1204	1.3404	1.5018	15
46	.66610	.74586	.89306	.1197	.3407	.5013	14
47	.66631	.74567	.89358	.1191	.3411	.5008	13
48	.66653	.74548	.89410	.1184	.3414	.5003	12
49	.66675	.74528	.89463	.1178	.3418	.4998	11
50	.66697	.74509	.89515	1.1171	1.3421	1.4993	10
51	.66718	.74489	.89567	.1165	.3425	.4988	9
52	.66740	.74470	.89620	.1158	.3428	.4983	8
53	.66762	.74450	.89672	.1152	.3432	.4979	7
54	.66783	.74431	.89725	.1145	.3435	.4974	6
55	.66805	.74412	.89777	1.1139	1.3439	1.4969	5
56	.66826	.74392	.89830	.1132	.3442	.4964	4
57	.66848	.74373	.89882	.1126	.3446	.4959	3
58	.66870	.74353	.89935	.1119	.3449	.4954	2
59	.66891	.74334	.89988	.1113	.3453	.4949	1
60	.66913	.74314	.90040	1.1106	1.3456	1.4945	0
M	Cosine	Sine	Cotan.	Tan.	Cosec.	Secant	M

48°

42°

M	Sine	Cosine	Tan.	Cotan.	Secant	Cosec.	M
0	.66913	.74314	.90040	1.1106	1.3456	1.4945	60
1	.66935	.74295	.90093	.1100	.3460	.4940	59
2	.66956	.74275	.90146	.1093	.3463	.4935	58
3	.66978	.74256	.90198	.1086	.3467	.4930	57
4	.66999	.74236	.90251	.1080	.3470	.4925	56
5	.67021	.74217	.90304	1.1074	1.3474	1.4921	55
6	.67043	.74197	.90357	.1067	.3477	.4916	54
7	.67064	.74178	.90410	.1061	.3481	.4911	53
8	.67086	.74158	.90463	.1054	.3485	.4906	52
9	.67107	.74139	.90515	.1048	.3488	.4901	51
10	.67129	.74119	.90568	1.1041	1.3492	1.4897	50
11	.67150	.74100	.90621	.1035	.3495	.4892	49
12	.67172	.74080	.90674	.1028	.3499	.4887	48
13	.67194	.74061	.90727	.1022	.3502	.4882	47
14	.67215	.74041	.90780	.1015	.3506	.4877	46
15	.67237	.74022	.90834	1.1009	1.3509	1.4873	45
16	.67258	.74002	.90887	.1003	.3513	.4868	44
17	.67280	.73983	.90940	.0996	.3517	.4863	43
18	.67301	.73963	.90993	.0990	.3520	.4858	42
19	.67323	.73943	.91046	.0983	.3524	.4854	41
20	.67344	.73924	.91099	1.0977	1.3527	1.4849	40
21	.67366	.73904	.91153	.0971	.3531	.4844	39
22	.67387	.73885	.91206	.0964	.3534	.4839	38
23	.67409	.73865	.91259	.0958	.3538	.4835	37
24	.67430	.73845	.91312	.0951	.3542	.4830	36
25	.67452	.73826	.91366	1.0945	1.3545	1.4825	35
26	.67473	.73806	.91419	.0939	.3549	.4821	34
27	.67495	.73787	.91473	.0932	.3552	.4816	33
28	.67516	.73767	.91526	.0926	.3556	.4811	32
29	.67537	.73747	.91580	.0919	.3560	.4806	31
30	.67559	.73728	.91633	1.0913	1.3563	1.4802	30
31	.67580	.73708	.91687	.0907	.3567	.4797	29
32	.67602	.73688	.91740	.0900	.3571	.4792	28
33	.67623	.73669	.91794	.0894	.3574	.4788	27
34	.67645	.73649	.91847	.0888	.3578	.4783	26
35	.67666	.73629	.91901	1.0881	1.3581	1.4778	25
36	.67688	.73610	.91955	.0875	.3585	.4774	24
37	.67709	.73590	.92008	.0868	.3589	.4769	23
38	.67730	.73570	.92062	.0862	.3592	.4764	22
39	.67752	.73551	.92116	.0856	.3596	.4760	21
40	.67773	.73531	.92170	1.0849	1.3600	1.4755	20
41	.67794	.73511	.92223	.0843	.3603	.4750	19
42	.67816	.73491	.92277	.0837	.3607	.4746	18
43	.67837	.73472	.92331	.0830	.3611	.4741	17
44	.67859	.73452	.92385	.0824	.3614	.4736	16
45	.67880	.73432	.92439	1.0818	1.3618	1.4732	15
46	.67901	.73412	.92493	.0812	.3622	.4727	14
47	.67923	.73393	.92547	.0805	.3625	.4723	13
48	.67944	.73373	.92601	.0799	.3629	.4718	12
49	.67965	.73353	.92655	.0793	.3633	.4713	11
50	.67987	.73333	.92709	1.0786	1.3636	1.4709	10
51	.68008	.73314	.92763	.0780	.3640	.4704	9
52	.68029	.73294	.92817	.0774	.3644	.4699	8
53	.68051	.73274	.92871	.0767	.3647	.4695	7
54	.68072	.73254	.92926	.0761	.3651	.4690	6
55	.68093	.73234	.92980	1.0755	1.3655	1.4686	5
56	.68115	.73215	.93034	.0749	.3658	.4681	4
57	.68136	.73195	.93088	.0742	.3662	.4676	3
58	.68157	.73175	.93143	.0736	.3666	.4672	2
59	.68178	.73155	.93197	.0730	.3669	.4667	1
60	.68200	.73135	.93251	1.0724	1.3673	1.4663	0
M	Cosine	Sine	Cotan.	Tan.	Cosec.	Secant	M

47°

43°

M	Sine	Cosine	Tan.	Cotan.	Secant	Cosec.	M
0	.68200	.73135	.93251	1.0724	1.3673	1.4663	60
1	.68221	.73115	.93306	.0717	.3677	.4658	59
2	.68242	.73096	.93360	.0711	.3681	.4654	58
3	.68264	.73076	.93415	.0705	.3684	.4649	57
4	.68285	.73056	.93469	.0699	.3688	.4644	56
5	.68306	.73036	.93524	1.0692	1.3692	1.4640	55
6	.68327	.73016	.93578	.0686	.3695	.4635	54
7	.68349	.72996	.93633	.0680	.3699	.4631	53
8	.68370	.72976	.93687	.0674	.3703	.4626	52
9	.68391	.72956	.93742	.0667	.3707	.4622	51
10	.68412	.72937	.93797	1.0661	1.3710	1.4617	50
11	.68433	.72917	.93851	.0655	.3714	.4613	49
12	.68455	.72897	.93906	.0649	.3718	.4608	48
13	.68476	.72877	.93961	.0643	.3722	.4604	47
14	.68497	.72857	.94016	.0636	.3725	.4599	46
15	.68518	.72837	.94071	1.0630	1.3729	1.4595	45
16	.68539	.72817	.94125	.0624	.3733	.4590	44
17	.68561	.72797	.94180	.0618	.3737	.4586	43
18	.68582	.72777	.94235	.0612	.3740	.4581	42
19	.68603	.72757	.94290	.0605	.3744	.4577	41
20	.68624	.72737	.94345	1.0599	1.3748	1.4572	40
21	.68645	.72717	.94400	.0593	.3752	.4568	39
22	.68666	.72697	.94455	.0587	.3756	.4563	38
23	.68688	.72677	.94510	.0581	.3759	.4559	37
24	.68709	.72657	.94565	.0575	.3763	.4554	36
25	.68730	.72637	.94620	1.0568	1.3767	1.4550	35
26	.68751	.72617	.94675	.0562	.3771	.4545	34
27	.68772	.72597	.94731	.0556	.3774	.4541	33
28	.68793	.72577	.94786	.0550	.3778	.4536	32
29	.68814	.72557	.94841	.0544	.3782	.4532	31
30	.68835	.72537	.94896	1.0538	1.3786	1.4527	30
31	.68856	.72517	.94952	.0532	.3790	.4523	29
32	.68878	.72497	.95007	.0525	.3794	.4518	28
33	.68899	.72477	.95062	.0519	.3797	.4514	27
34	.68920	.72457	.95118	.0513	.3801	.4510	26
35	.68941	.72437	.95173	1.0507	1.3805	1.4505	25
36	.68962	.72417	.95229	.0501	.3809	.4501	24
37	.68983	.72397	.95284	.0495	.3813	.4496	23
38	.69004	.72377	.95340	.0489	.3816	.4492	22
39	.69025	.72357	.95395	.0483	.3820	.4487	21
40	.69046	.72337	.95451	1.0476	1.3824	1.4483	20
41	.69067	.72317	.95506	.0470	.3828	.4479	19
42	.69088	.72297	.95562	.0464	.3832	.4474	18
43	.69109	.72277	.95618	.0458	.3836	.4470	17
44	.69130	.72256	.95673	.0452	.3839	.4465	16
45	.69151	.72236	.95729	1.0446	1.3843	1.4461	15
46	.69172	.72216	.95785	.0440	.3847	.4457	14
47	.69193	.72196	.95841	.0434	.3851	.4452	13
48	.69214	.72176	.95896	.0428	.3855	.4448	12
49	.69235	.72156	.95952	.0422	.3859	.4443	11
50	.69256	.72136	.96008	1.0416	1.3863	1.4439	10
51	.69277	.72115	.96064	.0410	.3867	.4435	9
52	.69298	.72095	.96120	.0404	.3870	.4430	8
53	.69319	.72075	.96176	.0397	.3874	.4426	7
54	.69340	.72055	.96232	.0391	.3878	.4422	6
55	.69361	.72035	.96288	1.0385	1.3882	1.4417	5
56	.69382	.72015	.96344	.0379	.3886	.4413	4
57	.69403	.71994	.96400	.0373	.3890	.4408	3
58	.69424	.71974	.96456	.0367	.3894	.4404	2
59	.69445	.71954	.96513	.0361	.3898	.4400	1
60	.69466	.71934	.96569	1.0355	1.3902	1.4395	0
M	Cosine	Sine	Cotan.	Tan.	Cosec.	Secant	M

46°

44°

M	Sine	Cosine	Tan.	Cotan.	Secant	Cosec.	M
0	.69466	.71934	.96569	1.0355	1.3902	1.4395	60
1	.69487	.71914	.96625	.0349	.3905	.4391	59
2	.69508	.71893	.96681	.0343	.3909	.4387	58
3	.69528	.71873	.96738	.0337	.3913	.4382	57
4	.69549	.71853	.96794	.0331	.3917	.4378	56
5	.69570	.71833	.96850	1.0325	1.3921	1.4374	55
6	.69591	.71813	.96907	.0319	.3925	.4370	54
7	.69612	.71792	.96963	.0313	.3929	.4365	53
8	.69633	.71772	.97020	.0307	.3933	.4361	52
9	.69654	.71752	.97076	.0301	.3937	.4357	51
10	.69675	.71732	.97133	1.0295	1.3941	1.4352	50
11	.69696	.71711	.97189	.0289	.3945	.4348	49
12	.69716	.71691	.97246	.0283	.3949	.4344	48
13	.69737	.71671	.97302	.0277	.3953	.4339	47
14	.69758	.71650	.97359	.0271	.3957	.4335	46
15	.69779	.71630	.97416	1.0265	1.3960	1.4331	45
16	.69800	.71610	.97472	.0259	.3964	.4327	44
17	.69821	.71589	.97529	.0253	.3968	.4322	43
18	.69841	.71569	.97586	.0247	.3972	.4318	42
19	.69862	.71549	.97643	.0241	.3976	.4314	41
20	.69883	.71529	.97700	1.0235	1.3980	1.4310	40
21	.69904	.71508	.97756	.0229	.3984	.4305	39
22	.69925	.71488	.97813	.0223	.3988	.4301	38
23	.69945	.71468	.97870	.0218	.3992	.4297	37
24	.69966	.71447	.97927	.0212	.3996	.4292	36
25	.69987	.71427	.97984	1.0206	1.4000	1.4288	35
26	.70008	.71406	.98041	.0200	.4004	.4284	34
27	.70029	.71386	.98098	.0194	.4008	.4280	33
28	.70049	.71366	.98155	.0188	.4012	.4276	32
29	.70070	.71345	.98212	.0182	.4016	.4271	31
30	.70091	.71325	.98270	1.0176	1.4020	1.4267	30
31	.70112	.71305	.98327	.0170	.4024	.4263	29
32	.70132	.71284	.98384	.0164	.4028	.4259	28
33	.70153	.71264	.98441	.0158	.4032	.4254	27
34	.70174	.71243	.98499	.0152	.4036	.4250	26
35	.70194	.71223	.98556	1.0146	1.4040	1.4246	25
36	.70215	.71203	.98613	.0141	.4044	.4242	24
37	.70236	.71182	.98671	.0135	.4048	.4238	23
38	.70257	.71162	.98728	.0129	.4052	.4233	22
39	.70277	.71141	.98786	.0123	.4056	.4229	21
40	.70298	.71121	.98843	1.0117	1.4060	1.4225	20
41	.70319	.71100	.98901	.0111	.4065	.4221	19
42	.70339	.71080	.98958	.0105	.4069	.4217	18
43	.70360	.71059	.99016	.0099	.4073	.4212	17
44	.70381	.71039	.99073	.0093	.4077	.4208	16
45	.70401	.71018	.99131	1.0088	1.4081	1.4204	15
46	.70422	.70998	.99189	.0082	.4085	.4200	14
47	.70443	.70977	.99246	.0076	.4089	.4196	13
48	.70463	.70957	.99304	.0070	.4093	.4192	12
49	.70484	.70936	.99362	.0064	.4097	.4188	11
50	.70505	.70916	.99420	1.0058	1.4101	1.4183	10
51	.70525	.70895	.99478	.0052	.4105	.4179	9
52	.70546	.70875	.99536	.0047	.4109	.4175	8
53	.70566	.70854	.99593	.0041	.4113	.4171	7
54	.70587	.70834	.99651	.0035	.4117	.4167	6
55	.70608	.70813	.99709	1.0029	1.4122	1.4163	5
56	.70628	.70793	.99767	.0023	.4126	.4159	4
57	.70649	.70772	.99826	.0017	.4130	.4154	3
58	.70669	.70752	.99884	.0012	.4134	.4150	2
59	.70690	.70731	.99942	.0006	.4138	.4146	1
60	.70711	.70711	1.0000	1.0000	1.4142	1.4142	0
M	Cosine	Sine	Cotan.	Tan.	Cosec.	Secant	M

45°

HOW TO MAKE 90° BENDS USING HAND BENDER

The take up of any ninety degree bend is the same as its radius. Hand bends made from pipe size 1/2" can be bent on a two inch radius, therefore it takes 3 1/8" of pipe to make the actual bend. This is found by multiplying the sine of one degree times ninety degrees times the radius of the bend.

When making a ninety degree bend to specific dimensions, follow this procedure:

1. From the end of a length of pipe, measure and place a mark representing the center to end dimension less the take up of one 90° bend. This will be the point where the bend starts and is called a bending mark.

2. From this point, measure and mark the amount of pipe required in the actual bend. This is also a bending mark. The bend will be made between these two marks.

3. Mark the center to end dimension (less take up of one 90° bend) of the other end of the bend from this mark and cut the pipe off at this point. Thread both ends.

4. Place the dimensioned pipe in a vise with one bending mark even with the edge of the vise jaw.

5. Align one end of the hand bender with the other bending mark and make the bend.

6. Remove bend from vise and check the degree of turn and the center to end dimensions.

The chart below is used in making bends from small size pipe.

PIPE SIZE	PIPE REQUIRED FOR 90° BEND	RADIUS OF BEND
1/2"	3 1/8"	2"
3/4"	4 3/4"	3"
1"	7 7/8"	5"

HOW TO MARK PIPE FOR 90° HAND BEND

Example Problem:

Mark off 1/2" pipe to conform to the dimensions of the 90° bend at right.

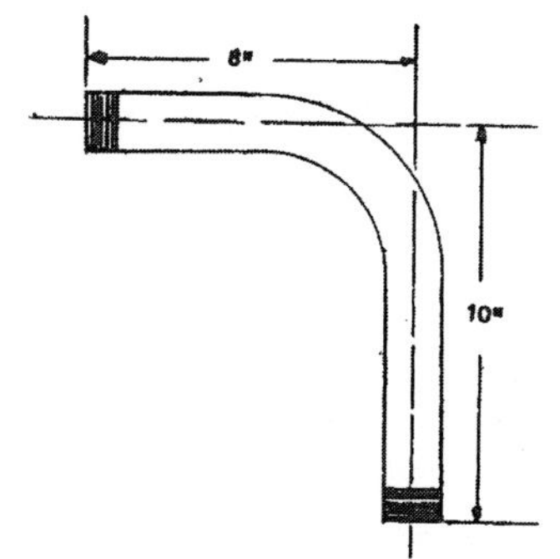

A = Center to end dimension minus radius of one 90° bend. (Eight inches minus two inches equals six inches).

B = Actual amount of straight pipe required to make the bend. (Sine of one degree times ninety degrees times radius).

C = Center to end dimension of the other end of the bend minus radius of one 90° bend. (Ten inches minus two inches equals eight inches).

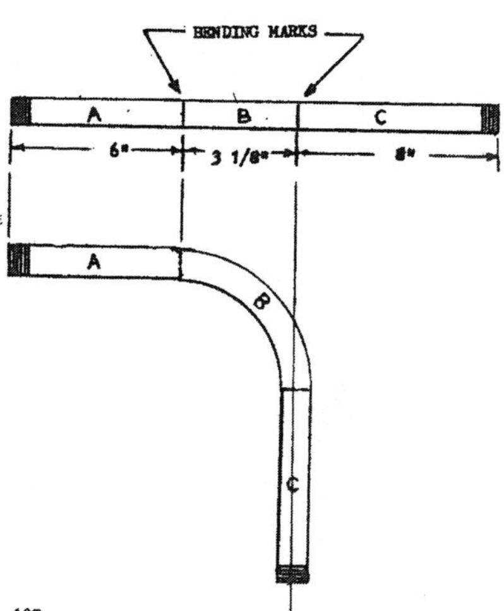

HOW TO LAY OFF DOUBLE 90 BENDS

On page 106 it was stated that the take up of one 90° bend is subtracted from the center to end dimension of a 90° bend. When marked on the pipe, this is the point for the start of the bend and is followed by the amount of pipe that it takes to make the actual turn.

When laying off a double 90° offset, the only difference is that the take up of two 90° bends is subtracted from the center to center measurement and this distance is marked off as straight pipe between the two bends.

Example Problem:

Mark off 1/2" pipe to conform to the dimensions of the offset at right.

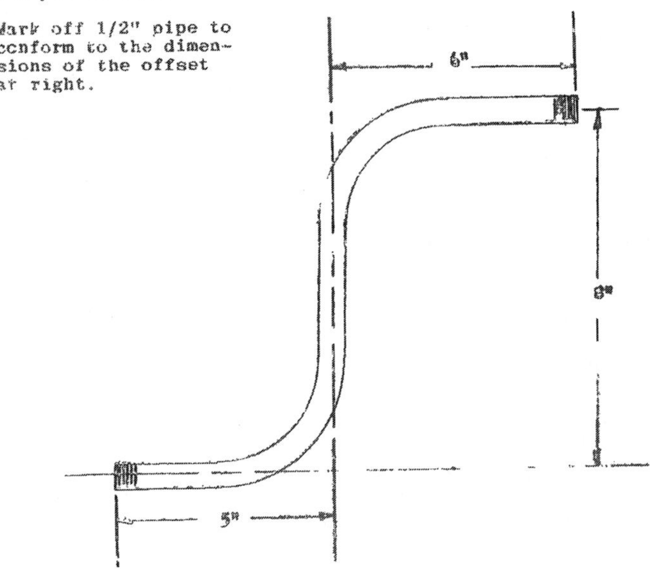

Pipe Size	Pipe Required For 90° Bend	Take Up of Bend
1/2"	3 1/8"	2"

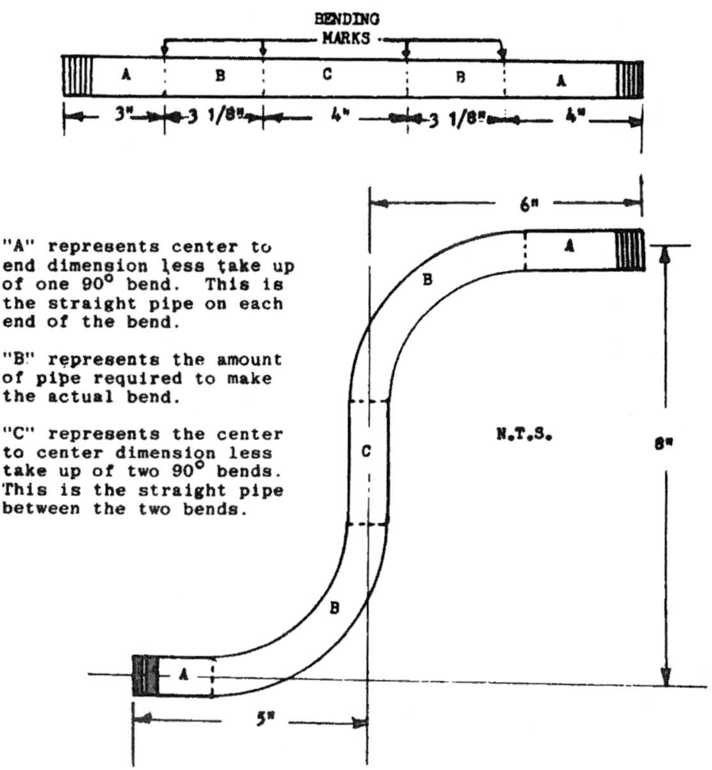

"A" represents center to end dimension less take up of one 90° bend. This is the straight pipe on each end of the bend.

"B" represents the amount of pipe required to make the actual bend.

"C" represents the center to center dimension less take up of two 90° bends. This is the straight pipe between the two bends.

After the bending marks have been laid off on the pipe, the bends can be made starting from either end of the pipe.